William Bezly Thorne

The Schott methods of the treatment of chronic diseases of the

heart

William Bezly Thorne

The Schott methods of the treatment of chronic diseases of the heart

ISBN/EAN: 9783337729332

Printed in Europe, USA, Canada, Australia, Japan

Cover: Foto ©berggeist007 / pixelio.de

More available books at **www.hansebooks.com**

THE
SCHOTT METHODS

OF THE

TREATMENT

OF

CHRONIC DISEASES OF THE HEART

WITH

AN ACCOUNT OF THE NAUHEIM BATHS, AND OF THE THERAPEUTIC EXERCISES

ILLUSTRATED

BY

W. BEZLY THORNE, M.D., M.R.C.P.

THIRD EDITION

LONDON
J. & A. CHURCHILL
7 GREAT MARLBOROUGH STREET
1899

PREFACE TO THE THIRD EDITION.

THE clinical experience which has been gained since the former editions of this work were issued has strengthened the conviction that the methods which it was their object to set forth are, as much as ever, deserving of the thoughtful attention of the medical profession. In the third edition old material has been rearranged, and new material introduced, in such a manner as to present two additional chapters, of which one deals more especially with conditions which should influence and modify the details of procedure, while the other offers instances of their application, and calls attention to features scarcely alluded to in the earlier issues, more especially to the institution of vascular repair. At the risk of some redundancy the cases originally reported have been retained in the concluding chapter.

W. BEZLY THORNE.

53, UPPER BROOK STREET,
  LONDON, W.,
    August, 1899.

PREFACE TO THE SECOND EDITION.

IN issuing the second edition of the only volume which, up to the present time, has presented, in concrete form, the system of cardio-therapy which August and Theodor Schott have evolved from principles first proclaimed by Stokes, I desire to express my thanks for the generous appreciation with which the first edition has been received, alike by friends and strangers in the profession of Medicine, and also the deep satisfaction with which it is now possible to view the increasing favour which is accorded to a novel, but potent, therapeutic expedient by those whose mission it is to relieve disease and suffering. The few months which have passed since these pages were first given to the medical profession have not diminished the confidence in the methods of which they treat, which then emboldened me to advocate a new departure. Happily, what are known as the Schott Methods promise soon to become an acknowledged and accepted medical practice.

W. BEZLY THORNE.

53, UPPER BROOK STREET,
    LONDON, W.,
        February, 1896.

PREFACE TO THE FIRST EDITION.

IN March, 1891, I was enabled, by the courtesy of the Editors of the *Lancet*, to lay before the medical profession a paper by Dr. Theodor Schott, in which were explained the principles and practice of the treatment of chronic diseases of the heart by means of mineral baths and exercises, which had been elaborated by him and his deceased brother. As judged by results, it attracted no notice and the system remained an unknown art in this country. In the early part of the year 1894 I was favoured with a similar opportunity of bringing forward a brief account of my own experience of the Schott system. Since that time I have received so many requests for further and more detailed information, that I am encouraged to meet an increasing demand by the publication of the following pages. They do not pretend to offer a complete or exhaustive exposition either of the science and art of the physical treatment of heart affections, or of the range of its application. That their scope is mainly limited by my own knowledge and observation, is my apology for defects which are only too manifest to myself.

W. BEZLY THORNE.

53, UPPER BROOK STREET,
  LONDON, W.,
    March, 1895.

CONTENTS.

CHAP.	PAGE
I. NAUHEIM AND ITS WATERS	11
II. BATHS	17
III. THERAPEUTIC MOVEMENTS	27
IV. CONDITIONS WHICH GOVERN THE APPLICATION OF THE METHODS	38
V. CONDITIONS TO WHICH THE METHODS ARE APPLICABLE	46
VI. CONDITIONS NOT PRIMARILY CARDIAC TO WHICH THE METHODS ARE APPLICABLE	82
VII. THE EXERCISES, DETAILED DESCRIPTION AND ILLUSTRATIONS	91
VIII. ILLUSTRATIVE CASES WITH DIAGRAMS	116

THE SCHOTT METHODS

OF THE TREATMENT OF

CHRONIC DISEASES OF THE HEART.

CHAPTER I.

BAD-NAUHEIM AND ITS WATERS.

SITUATED at the north-eastern extremity of the Taunus range, Nauheim—or, to give it its full name, Bad-Nauheim, the birthplace and headquarters of the treatment about to be considered—lies mainly on the gentle slope, which, looking south south-east, forms the foot of the Johannesberg.

The underground streams which have been brought into requisition for therapeutic drinking and bathing, have been tapped in the lowest part of the township—namely, at some little distance on either side of the stream which divides the park into two unequal portions; and it may be said at once that they rise from so great a depth as to preclude the possibility of subterranean communication with that small river.

The following analyses have been compiled from the observations recorded by Beneke, Prof. Will, of Giessen, Doctors August and Theodor Schott, and Dr. Uloth:—

ANALYSIS OF THE NAUHEIM WATERS.

Quantitative analyses of the Nauheim Waters, by Prof. WILL, Giessen, including the results of analyses by Drs. AUGUST and THEODOR SCHOTT, made especially to ascertain the several proportions of carbonic acid gas. The amounts of solids are given in grammes as contained in 1000 grammes of water.

CONSTITUENTS.	SPRING No. 12. Friedrich-Wilhelms-Quelle.	SPRING No. 7. Grosser Sprudel.	SPRING No. 11. Gas-Quelle.	KURBRUNNEN.	CARLS-BRUNNEN (Dr. Ulodj).
Chloride of Sodium	29·2940	21·8245	17·1888	15·4215	9·8600
,, ,, Lithium	0·0536	0·0492	0·0323	0·0267	Trace
,, ,, Potassium (Cæsium, Rubidium)	1·1194	0·4974	0·7174	0·5270	0·0726
,, ,, Ammonium	0·0712	0·0550	0·0433	0·0371	0·0113
,, ,, Calcium	3·3249	1·7000	1·2598	1·0349	1·0575
,, ,, Magnesium	0·5255	0·4402	0·3682	0·7387	0·2040
Bromide of ,,	0·0083	0·0060	0·0046	0·0063	0·0014
Iodide ,, ,,	Trace	—	—	—	—
Sulphate of Calcium	0·0352	0·0347	0·0190	0·0238	0·2277
,, ,, Strontium (with Baryta)	0·0499	0·0390	0·0403	0·0324	0·0087
Bicarbonate of Calcium	2·6012	2·3541	2·1473	1·1461	0·9515
,, ,, Iron	0·0484	0·0383	0·0313	0·0262	0·0147
,, ,, Manganese	0·0069	0·0065	0·0050	0·0080	Trace
,, ,, Zinc	0·0089	0·0104	0·0076	0·0070	Trace
Silicic Acid	0·0213	0·0325	0·0190	0·0186	0·0087
Arseniate of Iron	0·0002	0·00036	0·0005	0·00016	Trace
Phosphate ,, ,,	0·0007	0·00046	Uncertain	0·00034	0·0002
Organic Substances	Trace	Trace	Trace	Trace	Trace
Amount of Solid Constituents	35·3573	26·3539	21·1663	18·6935	12·4183
Specific Gravity	1·02757	1·02088	1·01685	1·01475	1·0089
Free Carbonic Acid Gas	*1·0074	†3·1756	1·4136	1·9622	1·4214
Temperature (Fahrenheit), Celsius	(95·54) 35·3	(88·88) 31·6	(81·68) 27·6	(70·52) 21·4	(59) 15
Outflow in 24 hours { Normal	1725	782			
in Cubic Meters { Valves half closed	782	529	28	16	6
Depth of Well in Meters	180	159·5			

* = 578·93 C.cm.   † = 1340·46 C.cm.

QUANTITATIVE ANALYSIS OF THE LUDWIGS SPRING.

By Prof. Will.

Results given in grammes in 1000 grammes of water.

Bicarbonate of Calcium	0·369
,, ,, Iron	0·009
,, ,, Manganese	—
,, ,, Magnesium	0·113
,, ,, Sodium	0·172
Sulphate of Calcium	0·028
Chloride of Potassium	—
,, ,, Sodium	0·341
,, ,, Calcium	—
,, ,, Lithium	0·001
Bromide of Sodium	—
Silica	0·012
Organic Substances	Trace
Total Solids	1·045
Carbonic Acid Gas contained under pressure of one atmosphere	1·254
Specific Gravity at temperature of 66·92° F. 19·40° C.	1·0010

QUANTITATIVE ANALYSIS OF THE SCHWALHEIM SPRING.

By Prof. von Liebig.

Results given in grains in one pound of water = Grammes in 1000 grammes of water.

Chloride of Sodium	11·9465 gr.	= 1·7060
Sulphate ,, ,,	0·6215 ,,	= 0·0900
Chloride of Magnesium	1·0826 ,,	= 0·1546
Carbonate of ,,	0·4185 ,,	= 0·0600
,, ,, Calcium	4·3140 ,,	= 0·6160
,, ,, Iron	0·0878 ,,	= 0·0125
Silica	0·1489 ,,	= 0·0212
Total amount of Solid Constituents	18·6188 ,,	= 2·6603
Amount of Carbonic Acid Gas held in solution at the pressure of one atmosphere	22·7258 ,,	= 3·2465
Specific Gravity	1·0022	
Temperature	51·8° F. 11·0° C.	

The springs which are used for bathing purposes are No. 12, No. 7, and No. 11; those employed for drinking purposes, the Kurbrunnen and the Carlsbrunnen, the Ludwigsbrunnen and the Schwalheimerbrunnen—mainly the former two. It will be observed that the bathing waters are endowed by nature with temperatures which suit them admirably to the purpose. As a matter of fact, it is only in exceptional cases that the waters have to be either artificially heated or cooled by ice.

A course of baths generally commences with the waters of the great Sprudel (thermal bath), freed from more or less of their natural gas, but, in any case, to such an extent as to induce a deposit of peroxide of iron and calcium carbonate, which, floating in the water, produces an opaque yellow coloration. To these, after a time, in increasing portions, are added one, two, three or even more litres of Mutterlauge—the uncrystallisable mother-liquor or waste product of the neighbouring works which provide large quantities of salt for the table. It is rich in chloride of calcium, and bromine. The smallest quantity, carried to the tongue with the tip of the finger, produces an intense burning suggestive of vesication. Next in order comes the Sprudel bath drawn from No. 7 or No. 12, according to the temperature desired, containing a residue of natural gas sufficient to retain the whole of the iron in solution, and to coat the body with unbroken relays of globules which, on the bather emerging from the water, are found to have produced, insensibly, a well-marked rubefacience and an agreeable glow of warmth. Then, finally, come the flowing Sprudel baths, probably the most powerful therapeutic baths known, in which the waters of either No. 7 or No. 12 forcibly enter and, through overflow pipes, leave the receptacle during the whole period of immersion. These, with their constantly rising and simmering globules, emerging from moving water of crystalline clearness, convey the impression of a bath of champagne, and induce a sense of exhilaration not unlike that which is associated with that favourite beverage. No patient, in any case, is allowed to take more than two, or three, or at the outside five, successive baths in as many days, a day of interval always being imposed. Where much infiltration or osteoid deposit has taken place, carefully regulated massage is made to succeed each bath.

Speaking generally, the effects of the baths are:—to lower the frequency and increase the force of the action of the heart, and to induce a sense of refreshment and invigoration which is shortly followed by an agreeable inclination to avail oneself of the hour's rest, in the recumbent position, which is enjoined as the invariable sequel. One of the more remote effects is to cause pain and even swelling of joints, and sometimes of nerve-sheaths, which have been previously affected by the gouty, rheumatic, or so-called rheumatoid processes. Such a condition generally endures for a few days only, but not only may it last longer, but it may be re-induced by each of the succeeding increments of balneological strength above mentioned. On the more permanent influences exercised in the circulatory and respiratory systems I shall enlarge in detail later on.

The range of morbid conditions which may be relieved by the internal administration and outward use of the Nauheim waters is very wide. They may be divided into those articular and numerous other changes which are dependent on the prolonged presence in the blood-stream of uric acid in excess, chronic affections of the heart and blood-vessels, with one notable exception; congestion of the abdominal and pelvic viscera; and the earlier stages of chronic affections, of the congestive or sub-inflammatory order, of the spinal nerve structures.

I propose, however, in this notice, to confine my observations to the systematic use of the saline baths and of regulated movements of the body in chronic affections of the heart, according to methods elaborated after years of careful study by Prof. Theodor Schott and his deceased brother, Dr. August Schott.

CHAPTER II.

BATHS.

IT has already been stated that the immediate effect of immersion in the Nauheim baths is to reduce the frequency and increase the force of the action of the heart. For example, at a time when my own pulse averaged 74 beats per minute in the recumbent, and 84 in the sitting position, the heart and vessels being sound, I found it, on four separate occasions, to have fallen, within two minutes of immersion in a Sprudel bath, to from 60 to 64. In ten minutes it had risen to from 66 to 68, and there remained during the period of immersion, which in no case exceeded fifteen minutes. The exertion of dressing raised it to from 76 to 78; but, after the prescribed recumbent position had been assumed, it returned to from 62 to 66, with increased volume, and so remained during the period of repose. It will, therefore, be observed that the influence of the bath was not limited to the period of immersion.

These observations were made in August, 1893, and were verified a year later in the course of a series of five-and-twenty baths. Dr. John Broadbent was present during the twelfth, and, having traced the area of cardiac dulness before and after immersion, certified to a recession, averaging one third of an inch, in the general outline traced from the sternal to the mammary region.

By way of comparison, the following case may be quoted. A patient, aged forty-six, whose health had been declining for several years, was found to have a pulse of 80 in the recumbent, and of 88 in the sitting, position. While he stood it varied from 100 to 104; and if he walked ten paces it rose to from 120 to 130. The apex was found to beat an inch outside the nipple line. Within two minutes of immersion in his first thermal bath (spring No. 7, divested of the greater part of its carbonic acid gas, temp. 90·5° F.) the pulse had fallen to 70, and, judged by the finger, appeared to have doubled in volume; at the end of four minutes it was 68; in six minutes 66; in eight minutes 68; and while standing, after dressing, it was 90. Before he left the bath, after an immersion of ten minutes, the apex beat was found to have receded half an inch in the direction of the mesial line; and nails and fingers, which had been snow-white up to the junction of the second with the first phalanx, had assumed a healthy flesh tint.

The immediate effect of the first few baths is to produce a sense of oppression at the præcordium, under the influence of which the patient breathes slowly and deeply for two or three minutes. Respiration then becomes easy and continues slower by from two to four breaths a minute.

The effect on the peripheral vessels is to increase their carrying power. A glowing sense of warmth is experienced in the extremities and in the surface of the body generally. The veins are stimulated to a similar activity. In fact, the general arterial capacity, systemic and pulmonary, is increased, and, without loss of blood, the relief of a general bleeding is afforded to an overloaded and labouring heart.

Such being the results of a carefully graduated and regulated series of immersions in these saline waters, it can scarcely be matter for surprise that in three or four days, especially in cases in which the flow of urine has been scanty, there ensues a free diuresis which may continue for days or weeks; that metabolic change becomes accelerated and improved; that deeply-seated organs, more especially the liver and pelvic viscera, are relieved of congestion and partake in the general impulse to functional health; and that the heart, relieved of its burden, and contracting fully and without hurry on its contents, derives from an improved coronary circulation materials for the repair of its weakened or damaged tissues. It is suggested by Dr. Schott that these effects are produced partly by the cutaneous excitation induced by the mineral and gaseous constituents of the waters, and partly by a more prolonged stimulation of the nerves of sensation excited by imbibition into the superficial layer of the corium. According to this hypothesis, each sensitive nerve branch distributed over the surface that has been immersed transmits to its parent centre an influence which is centrifugally reflected to the vasomotor system and to the ganglia which control the action of the heart. That the nerve centres are brought under powerful influence is attested by the remarkable trophic changes which may be observed to follow a course of these baths, unaided by the internal use of mineral waters or pharmaceutical remedies, in cases of anæmia, wasting, neurasthenia, and, above all, in cases of osteoarthritis.

The rehabilitation of the trophic, and probably of other central nerve tissues, is so lasting that progressive improvement may be observed for three or four months after the completion of the course. It need hardly be pointed out that such a process of general health restoration is a factor of scarcely secondary importance in cases in which the condition of the heart presents the main indication for treatment.

It is necessary here to state that it is not claimed that these waters are unique in their therapeutic influences; on the contrary, from the earliest days, Prof. Schott and his brother have insisted that similar, if not indeed identical, effects may be derived from baths artificially prepared so as to resemble the Nauheim waters in their principal mineral ingredients. Thus it is recommended that the treatment should commence with a 1 per cent. solution of chloride of sodium, and that the strength should be gradually raised to 2 or 3 per cent. For increasing the cutaneous excitation, chloride of calcium is the salt to be relied on. The initial strength of the bath with regard to that ingredient should be 0·2 per cent., approximately that of No. 7 spring, and, by increasing additions, it may be raised to 0·3 per cent., that is to about the strength of No. 12 spring, and eventually to 0·5 per cent. Such varying degrees of concentration may be obtained by the proportional use of the crystallised Nauheim bath salt, of the mother-lye or of calcium chloride, and of Mediterranean sea residue which also contains chloride of sodium and of calcium in the proportions necessary for the preparation of initial baths, together with traces of bromides and iodides.

For the production of carbonic acid effervescence

the action of hydrochloric acid on bicarbonate of soda may be relied on. As a state of chemical purity is not required by the circumstances of the case, the articles of commerce are sufficient for the purpose. Two ways of employing the reagents are suggested, the one calculated to induce slow and gradual, the other rapid and almost immediate, effervescence. In the case of the first, the various salts, including the requisite proportion of bicarbonate of soda, having been dissolved, a bottle containing the acid is laid at the bottom of the bath, and the stopper having been withdrawn it is moved about from time to time. The bath will be ready for use in two or three hours. For the more rapid production of effervescence, the stopper of the bottle containing the acid is loosened, but retained in position; the bottle having then been inverted and lowered until its mouth is just below the surface of the water, the stopper is withdrawn, and the bottle is moved about so as to diffuse a layer of acid as uniformly as possible over the surface of the bath. By this means the bath will be prepared in about five minutes. It will be useful to employ baths of three degrees of effervescence:

Mild ... ... $\frac{1}{2}$ lb. $NaHCO_3$ to $\frac{3}{4}$ lb. HCl (25 per cent.)

Medium ... ... 1 lb. $NaHCO_3$ to $1\frac{1}{2}$ lb. HCl.

Strong (Sprudel strength) ... 2 lbs. $NaHCO_3$ to 3 lbs. HCl.

Except in the case of porcelain baths it is desirable to ensure a slight excess of alkali in order to prevent corrosion.

More convenient, however, for general use are Sandow's effervescing tablets, of which from four to eight, according to the strength desired, may be placed in the bath on either side of the place which bather will occupy.*

As evidence of the fact that the virtues which are claimed for the Nauheim waters are not peculiar to them, it may be interesting to record the result of some observations recently made on the effects of baths consisting of the waters of Llangammarch Wells, in Breconshire, the mineral constituents of which, according to an analysis made by Dr. Dupré in 1883, are as follows :—

		Parts per 1000.
Chloride of Sodium ...	189·56	2·708
Chloride of Calcium ...	84·56	1·208
Chloride of Magnesium	24·31	0·350
Chloride of Barium ...	6·26	0·090
Carbonate of Calcium	2·80	0·040
Silica	1·40	0·002

308·89 grains per gallon.

Iron Protoxide a trace.
Potassium
Strontium
Lithium   } Minute traces.
Bromine

Mr. S. A. Vasey, F.I.C., F.C.S., kindly permitted the first experiment to be made on himself, and the observations were made by Mr. Hugh Bennett, M.R.C.S., L.S.A., of Builth, and myself. The pulse, before immersion, was 100. From the second to the eighth minute the same oppression of the

* The agent for the sale in England of Dr. Sandow's preparations is Mr. M. Buchner, of 149, Houndsditch, London, E.C.

NOTE.—*Areas of cardiac dulness and apex beats indicated by red lines and crosses, respectively, refer to observations made after either baths or exercises.*

DIAGRAM A.

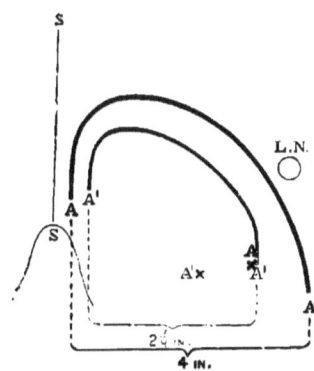

A A, area of cardiac dulness before immersion in bath of Llangammarch water. Temp. 92° F.
A¹ A¹. the same after ten minutes' immersion.
A×and A¹×. positions of apex beat at corresponding stages.
R.N. and L N., right and left nipples.
S S. mid-sternal line.

DIAGRAM B.

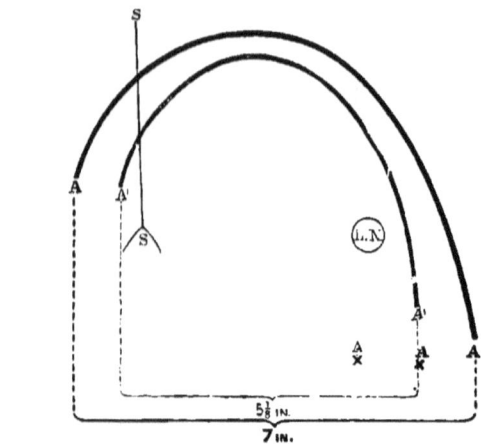

A A and A' A', before and after eight minutes' immersion in a Llan-
gammarch-Nauheim bath.
A × and A ×, positions of the apex beat at corresponding stages.
R.N. and L.N., right and left nipples.
S S mid-sternal line.

breathing which ensues on immersion in a Nauheim bath was experienced. At the eighth minute the pulse was found to be 78; after an immersion of ten minutes it was 76; after dressing 86, and an hour later 82. The change which was effected in the area of cardiac dulness is shown in the accompanying diagram (A). The temperature of the bath was 92° F.

The next observation was made on myself by Mr. Bennett. Before undressing the pulse was 96.

On immersion it fell at once to 80
At the end of the 1st minute it was 68
,, ,, ,, 2nd ,, ,, 69
,, ,, ,, 3rd ,, ,, 72
,, ,, ,, 4th ,, ,, 73
,, ,, ,, 5th ,, ,, 74
,, ,, ,, 6th ,, ,, 70
,, ,, ,, 7th ,, ,, 68
,, ,, ,, 8th ,, ,, 72
,, ,, ,, 9th ,, ,, 76
,, ,, ,, 10th ,, ,, 77

The recession of the area of cardiac dulness at the sternal end of the arc was ½ inch, and at the apex ⅜ inch. The temperature of the bath was 89° F.

A third observation was made on the following day by myself on a woman aged forty-seven, the subject of aortic stenosis, albuminuria, partial ascites, and œdema of the lower extremities. In this instance the strength of the bath, in chloride of sodium, was increased to that of the Nauheim spring No. 7. Before an immersion of ten minutes the pulse was 92, and after it 88; when the patient was partly dressed it was 84. The temperature of the bath was 92° F. The accompanying diagram (B) shows the diminution which was effected in the area of cardiac dulness.

In each of the above three cases an increase in the volume of the pulse similar to that effected by the Nauheim baths was found to take place, but the after-glow which followed immersion in the bath, which had been strengthened by the addition of chloride of sodium, appeared to be greater than in the other two instances. These observations tend to prove the correctness of Dr. Schott's disinterested contention that the Nauheim baths enjoy no monopoly of heart-therapy.

The following sphygmographic tracings and notes of pulse pressure recorded by the sphygmomanometer, are borrowed from a paper of Dr. Theodor Schott's, which I was the means of laying before the medical profession in Britain in 1891,* and afford evidences, which might be indefinitely multiplied, of the invigorating influence which the baths exercise on the heart and the circulation.

Tracings taken from a patient aged thirty-one, suffering from cardiac weakness.

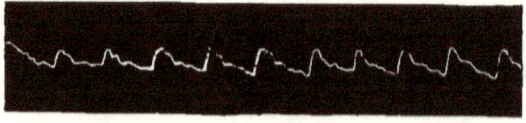

FIG. 1.

Before bathing: Frequency of the pulse, 94; pressure of the pulse, 120 millimetres of mercury.

* *Lancet*, May 23rd and 30th, 1891.

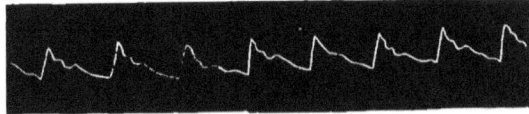

FIG. 2.

After the use of a Nauheim effervescent mineral bath of 87° F., duration fifteen minutes: Frequency of the pulse, 72; pressure of the pulse, 110 millimetres of mercury.

Tracings taken from a patient aged forty-six, affected with stenosis ostii arteriosi sinistri. Exercises were superadded to baths on the ninth day of treatment.

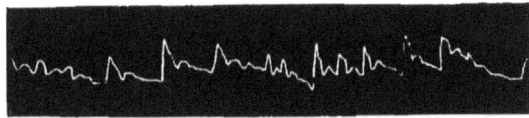

FIG. 3.

Before the beginning of the treatment: The pulse could not be counted (more than 150); pressure of the pulse, 82 millimetres of mercury.

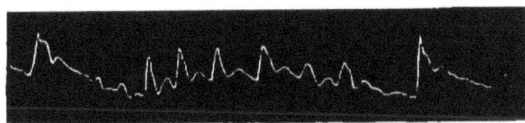

FIG. 4.

After the first bath, containing 1 per cent. of salt, temperature 89·5° F., duration ten minutes.

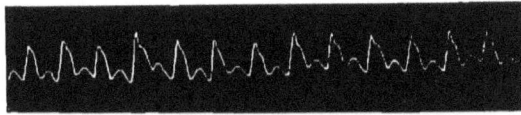

FIG. 5.

Eighth day of treatment by baths: Frequency of the pulse, 144; pressure of the pulse, 95 millimetres of mercury.

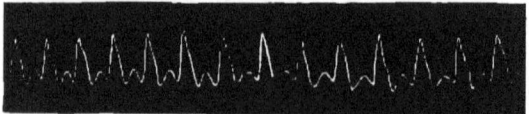

Fig. 6.

Ninth day of treatment, after half-an-hour's exercises with resistance: Pressure of the pulse, 110 millimetres of mercury.

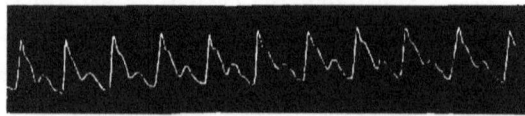

Fig. 7.

Fourteenth day of treatment: Frequency of the pulse, 108; pressure of the pulse, 115 millimetres of mercury.

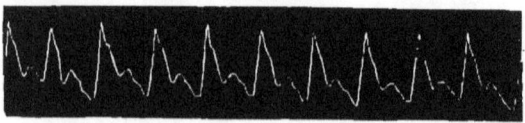

Fig. 8.

After three weeks' treatment: Frequency of the pulse, 108; pressure of the pulse, 125 millimetres of mercury.

CHAPTER III.

THERAPEUTIC MOVEMENTS.

THE treatment of cardiac affections, as practised by Prof. Schott, is not, however, limited by the therapeutic influences of the baths. As the result of a series of elaborate and prolonged experiments carried out by him and his deceased brother, Dr. August Schott, a system of exercises has been devised which yields results as striking as those of the baths. Their effect is illustrated by the following records, which were made in Nauheim by myself in conjunction with Dr. Hermann, of Charkoff, in August, 1893. The patient was a stout, well-built, fresh-looking man, forty years of age. He brought letters from Professor von Jürgensen of Tübingen and from his brother, who is a medical man, both of which described him as having been addicted to alcoholic excesses and being the subject of cor adiposum. Before the exercises the heart sounds were barely audible through a binaural stethoscope, and the apex beat was inappreciable. There was some œdema of the lower extremities.

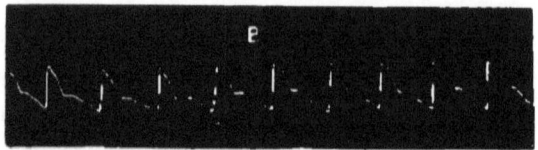

FIG. 10.
After twenty minutes' exercises.

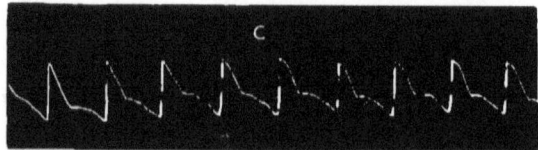

FIG. 11.
After thirty-five minutes' exercises.

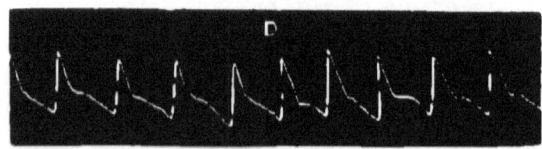

FIG. 12.
After forty-five minutes' exercises.

The following diagram (Fig. 13) gives, on a reduced

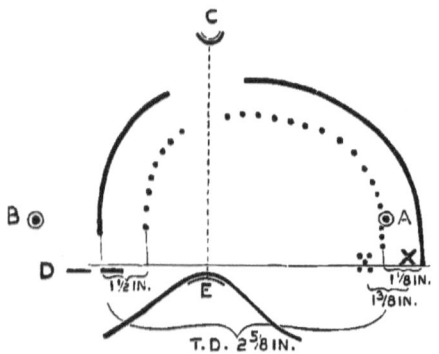

FIG. 13.

A, left nipple; B, right nipple; C, fossa jugularis; D, upper level of hepatic dulness: E, ensiform cartilage; T D, total diminution of area of cardiac dulness; ×, ⋮, positions of apex beat.

scale, the tracings of the area of cardiac dulness which were taken before, and at the conclusion of, the same exercises.

These exercises have been denominated by their inventors "Widerstandsgymnastik," or gymnastic with resistance. They may perhaps be more conveniently termed resisted movements or exercises. They consist of movements calculated to bring into successive and regulated action almost every collective system of voluntary muscles which is comprised in the human frame. Each succeeding movement is resisted by an attendant to such an extent as to oppose without arresting it. They consist of slowly-conducted flexion and extension, adduction and abduction, and rotation, in orderly succession, of the arms, the trunk, and the lower extremities. Each movement must be slowly and evenly made with a definite and uniform effort on the part of the patient. The office of the attendant is not limited to resisting the movements with equally uniform, but slightly inferior, force, but it is his duty to impose a short interval between each movement, to enjoin slow and regular breathing, and, more especially, by observing the rate of breathing and its force as indicated by the alæ nasi, to ensure that no undue strain is placed on the heart and lungs. He is also charged to guard the patient from perspiration and the slightest approach to palpitation of the heart. Either of these indications must be the signal for an interval of repose, during which the part being exercised is either left to hang at rest, or is supported by the hand of the attendant, who, under no circumstances, is allowed to grasp or in any manner constrict any portion of the patient's body. As the course proceeds, the energy of the movements, and consequently the force of the resistance, are gradually increased. Before and after the séance, and, if necessary, at some intermediate period, the physician maps out the dimensions of the heart by percussion and ascertains the state of the pulse, to satisfy himself that the effects of the exercises are satisfactory. The results, in fact, are such as would scarcely be believed by any but an eye-witness. It is by no means uncommon, in cases of dilatation, to see, within one hour, the oblique long diameter of the heart's area of dulness diminish by from three-quarters of an inch to an inch and a quarter, and, perhaps more surprising still, to observe a diminution by as many as two inches, in vertical measurement, of a liver which at first extended to the umbilical level; and to hear the patient, at the conclusion of what cannot be described as an ordeal, volunteer the statement that a load has been removed from the præcordium, that he breathes easier and more deeply, and experiences a sense of general relief.

It is not suggested that the whole of such a gain is permanent, for in the time that intervenes before the next day's exercises or bath, as the case may be, the dilated and congested organs tend to their former dimensions, but, be it well observed, they do not wholly relapse. On the contrary, each contraction ensures a proportional permanent gain, until, at the end of a few weeks, the attenuated and dilated heart and the congested liver may have recovered either their normal dimensions, or, at any rate, such contraction and compensatory power in the one case, and resolution in the other, as constitute them practically sound.

The resources of Prof. Schott and his brother did not come to an end with the conception of this system of physical treatment. With a view to enabling physicians and patients to maintain treatment of the same kind without dependence on a second person, they devised a method in the practice of which the patient is instructed, himself, to supply the necessary resistance or its equivalent, and by the aid of these self-resisted movements he is able to carry on and, from time to time if need be, resume a therapeutic process of unquestionable value.

In the course of his memorable lectures on surgical pathology, Sir James Paget quoted the profound observation of Treviranus that "each single part of the body, in respect of its nutrition, stands to the whole body in the relation of an excreted substance." In view of what may be achieved by means of the Schott system of therapeutic exercises, we may go further, and say that each part of the body, through its motor nerves, is capable of exercising a health-sustaining, and in some cases a health-restoring, influence on the heart and circulatory system, and consecutively on the entire organism. It has been shown that, even in health, the heart may present, under the alternating influences of exercise and repose, very appreciable variations in size; and Stokes, long ago, insisted that exertion, under suitable conditions, may promote the health of a damaged heart. On the other hand, there are not wanting examples of hearts that have been morbidly dilated, and to that extent damaged, either temporarily or permanently, by strains disproportionate to their strength. Drs. August and Theodor Schott enjoy the distinction, and are entitled to the credit, of having brought the physiological relations of exercise, function, and repair into obedience to a therapeutic system which yields results in the treatment of diseases of the heart hitherto unknown and unlooked for. Such service brings honour to their profession and deserves the gratitude of mankind.

What has been said of the influence of the baths

applies equally to the therapeutic exercises, except that retardation of the pulse is not so rapidly effected, even though its force and volume be manifestly increased; and that, in the nature of things, it would not be desirable to prolong immersion beyond fifteen or, even in cases of exceptional tolerance, twenty minutes; while, on the other hand, there are few patients who may not be kept under exercise for half an hour, and some can undergo an hour's treatment without fatigue. There are, therefore, in matters of secondary detail, differences between the baths and the exercises, and it rests with the physician to decide whether one should be brought into requisition to the exclusion of the other, or both be employed at suitable times and intervals.

As with the baths so with the exercises, therefore, the following immediate results may be looked for in the majority of patients afflicted with a damaged or weakened heart: retardation of the pulse and increase of its volume; contraction of the heart, generally first on the right side (it is rare for the left ventricle not to share in the contraction); slower and deeper breathing, with a sense of lightness and relief in the chest; a better colour of the lips and improved facial aspect; and, where that organ is congested, a notable diminution in the dimensions of the liver. Systematic administration of the exercises is generally followed in a few days by marked, and often long maintained, diuresis.

In the course of the first few movements a bruit, due to stenosis, may be observed to become accentuated; before the series has been completed murmurs, resulting from valvular incompetence other than that caused by actual lesion, may be diminished, then modified to duplication, and finally obliterated; heart sounds which were barely to be heard may become distinctly audible; and an apex beat that could not, under any circumstances, be detected may become appreciable to the touch.

The increase of the general arterial capacity is not less striking in the case of the exercises than of the baths. Within a few minutes the size of radial artery, as gauged by the touch, may seem to have doubled, and, before the series of movements has been completed, cheeks and fingers that were cold and either white or bluish-red, glow with warmth and healthy colour. The motor nerves, called into orderly, regulated, and, above all, not exhausting activity, seem to exercise centripetal and reflex influences similar to those which are brought into action by the baths through the nerves of sensation. The increased capacity of the vessels, down to the smallest capillaries, enables the heart so to contract as to empty its cavities at each stroke; while, at the same time, the ganglia, which control its action, seem to enforce a tonic contraction, which, renewed and maintained from day to day, leads to the establishment of a better habit of both function and repair.

It may be well, now, to gather together in brief summary the effects of these simple but remarkable movements, and I am favoured with permission to do so by quoting the following lines from an article by Sir Philip C. Smyly, which appeared in the *Dublin Journal of Medical Science* (September, 1894):—

"Take for example, the four following phenomena:—

1. The colour before the movements is a purple-blue in the cheeks and hands and feet.	1. The colour after twenty minutes or so becomes red, and the blue gradually disappears from the hands and feet.

2. The forehead, neck, and ears, etc., are a waxy white.

2. The forehead, neck, and ears, etc., become pink.

3. The pulse is rapid and blood-pressure low in the arteries.

3. The pulse slows and becomes full, the blood-pressure rises.

4. The area of dulness over the heart is large.

4. The area of dulness diminishes at times as much as an inch or more in diameter.

These results are due to:—

1. Increased arterial circulation, due to "the diminution of peripheral resistance." *
2. Diminished venous congestion, due to larger quantity of red blood in the arteries.
3. Diminished work for the heart, due to the free circulation of the blood in the arteries.

. . . . . there will ever be a feeling against this treatment until it is clearly seen and believed to be true:—

1. That the movements relieve the back pressure on the heart.
2. That the diminution in the size of the heart is due to the absence of excess of blood in its cavity.
3. That this is attained by there being more room in the arteries.
4. That the heart muscle gains strength by having room to contract.
5. That the contraction being more complete, it takes a longer time, thus making the pulse slower, and, at the same time, fuller.

---

* "Mechano-Therapy," A. Symons. Eccles. Pract. Aug., 1894, p. 114.

PLATE I.

To follow page 34.

PLATE II.

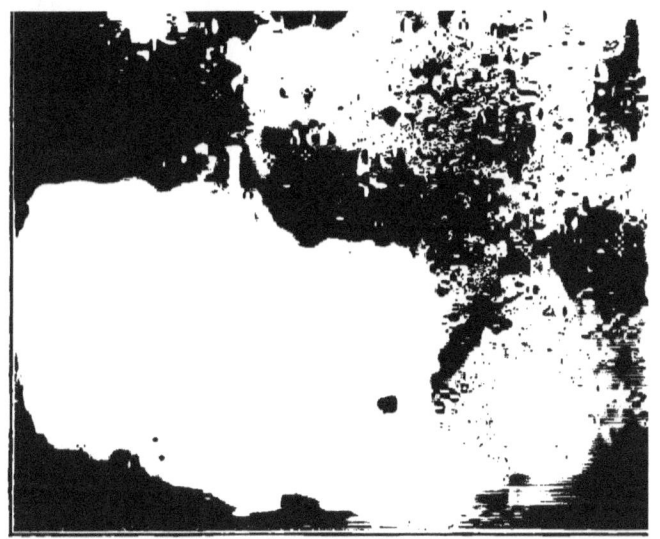

To face Plate I.

6. Being able to send on more blood it is ready to receive more, and thus removes venous congestion.
7. The strength gained by the heart is due to the freedom to contract fully."

Various suggestions and arguments have from time to time been advanced to show that the diminution of the cardiac dulness, referred to as following on the administration of both baths and exercises, either is imaginary, or, if not imaginary, indicative of no actual change in the volume of the heart itself. It has, for example, been asserted that belief in such modifications of the area of dulness must either be due to the influence of preconceived ideas or to defective methods of procedure. Such suggestions have, however, emanated from those whose practical acquaintance with the treatment has been very limited, and is contradicted by so overwhelming a consensus of opinion on the part of those who have made observations as extensive as they have been painstaking, that they need not be further alluded to.

A more plausible objection is that the diminution of the area of percussion-dulness must be attributed to an increase of pulmonary inflation, and of consequent overlapping of the heart. That such is not the case is shown by the fact that the position of the diaphragm either remains unchanged, or if it move, does so in an upward direction. Moreover, the interposition of a cushion of breathing lung of increased superficies and depth might be expected to muffle the force of the impact of the apex against the chest wall. The fact that migration in the direction of the mesial line is generally accompanied by accentuation, and in no case by diminution, of the impulse would appear to be of itself a sufficient answer.

The same reasons traverse the theory that the effect of baths and exercises is to induce rotation of the heart, on the basal vessels as a centre, in such a manner as to cause it to present a smaller area to percussion, apart from any actual change of bulk or outline. It is to be observed that no evidence, either actual or presumptive, has been adduced in support of that suggestion; whereas the results of clinical observation combine to contradict it. The descent of the apical portion of the heart could only take place in conjunction with a lowering of the average level of the diaphragm, and a corresponding rise of that part of the basal portion of the heart which is situated to the right of the great vessels. No such changes have ever been observed to take place. On the other hand, it is quite true that the subsidence of gastric dilatation may cause the heart to assume a lower and more normal position in the thorax; but, as could be shown by numerous records, the several regions of the heart, under those circumstances, maintain their relative positions to the vertical middle line of the chest. Happily, however, demonstration of the therapeutic value of these methods is not dependent on the measure of accuracy which may characterise either the theories or the clinical observations of those who carry them into practice. Their warranty is to be found in the life and health of those who have put them to the test, and who not long ago would have found themselves condemned to bear the burden of life on a downward course, with but partial and temporary alleviation, until the melancholy end had been reached.

Before leaving the subject of the cardiac dulness, it may be well to call attention to the following photographic reproductions of radiograms on a reduced scale. Those given in Plate I. are placed at my

PLATE III.

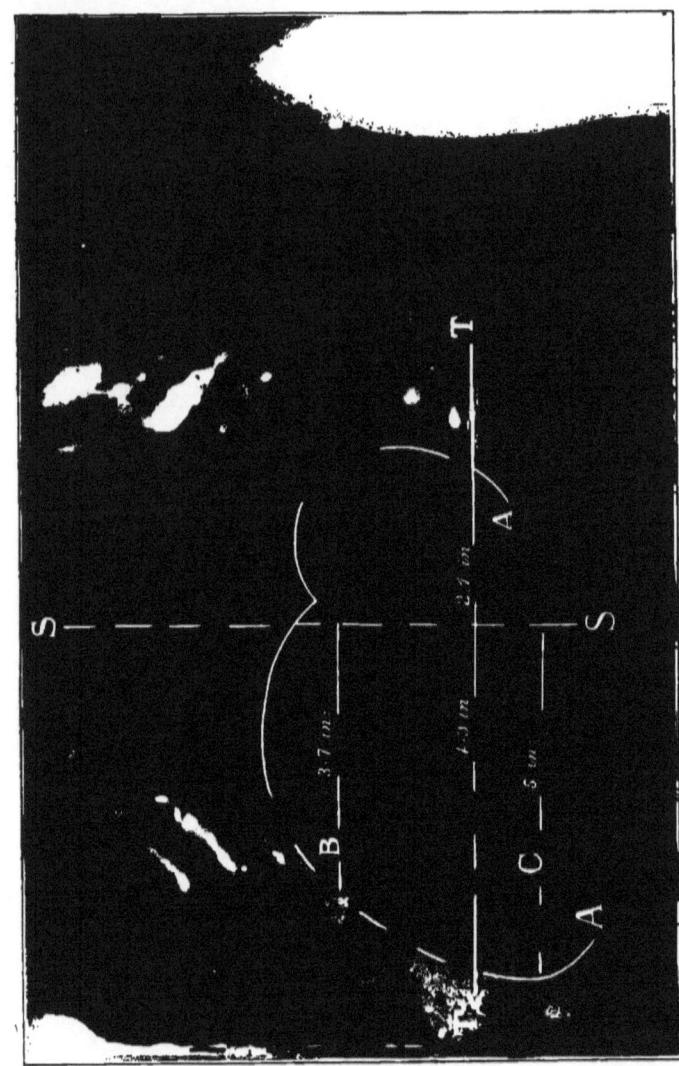

A A, percussion outline; T T, line drawn to intersect the 8th ribs at the same level in both radiograms; B B, line drawn 2 inches above T T for purpose of comparative measurement; C, line drawn through the promontory of the apex.

PLATE IV.

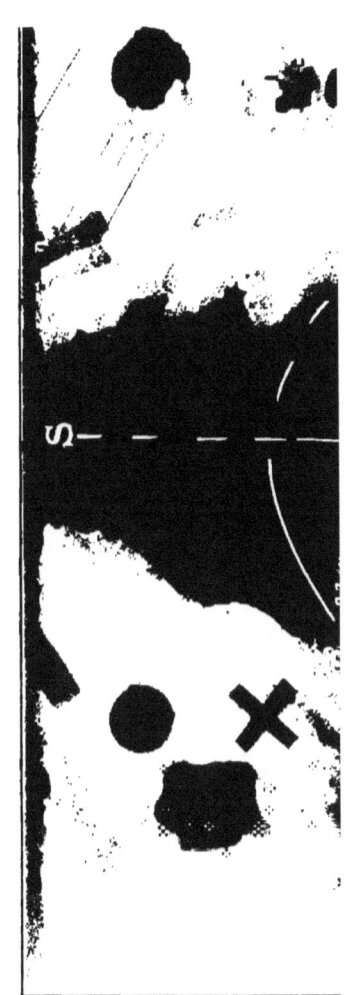

To face Plate III.

PLATE V.

The black lines were traced on transparent paper placed on the screen before, and the red lines after, the administration of seven lightly-resisted movements. The patient was between 50 and 60 years of age, was the subject of winter bronchitis, and had a weak and dilated heart.

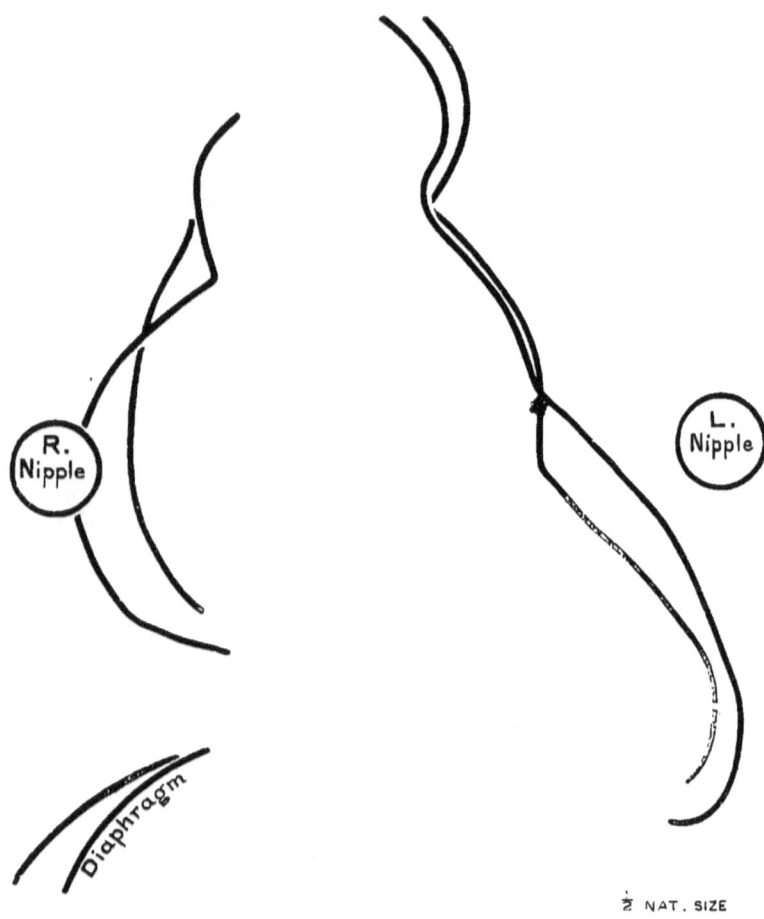

½ NAT. SIZE

To face page 37.

disposal by Prof. Schott. They were taken from a man, forty-four years of age, who, after rheumatic fever and in conjunction with alcoholic excesses, was found to be the subject of mitral insufficiency, dilatation of both ventricles (mainly of the right), commencing arterio-sclerosis, and angina pectoris. A course of exercise brought relief of all his symptoms. Measurement of the radiograms of natural size give the following results as following half-an-hour's resisted exercises :—

	Before.	After.
At the level of the third rib	15·6	14·2 centimetres.
Between the mamillæ	19·9	18·2 ,,

It is interesting to observe that, before the administration of the exercises, the leaden discs fixed to the mamillæ rose and fell, as the result of dyspnœa, to such a degree as to produce an oval shadow, whereas after their administration the movement was so much reduced that the same discs show a nearly circular outline. It is also noteworthy that, after the exercises, the diaphragm is found to have risen on both sides of the heart, and that there is no evidence of rotation of that organ on the basal vessels.

The reduced photographs given on Plates III. and IV., which have already been published,* were taken by Mr. Wilson Noble, in September, 1896, from a patient of Mr. Harris of Cranbrook, affected with mitral disease and œdema of the lower extremities. The white lines indicate the areas of dulness, as marked on the chest by Mr. Harris and myself, before and after the administration of seven lightly resisted movements.

The accompanying tracings on paper fixed to the fluorescent screen were made by Mr. Harris and myself, in June, 1898, with the assistance of Mr. Wilson Noble. (See Plate V.)

* "Cardio-vascular Repair." J. & A. Churchill.

CHAPTER IV.

CONDITIONS WHICH GOVERN THE APPLICATION
OF THE METHODS.

SPEAKING generally, the baths, as compared with the exercises, offer special advantages for the treatment of failure of compensation and of cardio-vascular degeneration, whether or no there be co-existing valvular lesions. But it should be pointed out that in some extreme cases in which either the cardiac condition itself or some dependent, or independently occurring, complication forbids the removal of the patient from his bed, the exercises, modified, if need be, to meet the requirements of the case, may prove to be of value in preparing the patient to undergo treatment by baths, more especially if they should not be easy of access. The exercises, therefore, may be at once brought into use, with or without the more powerful influences of the baths, to meet either failure of compensation or the general condition of those in whom degeneration has not made such advances as must in the nature of things be irreparable. The combined application of the two methods may, in some instances, be expected to yield results which could not be obtained by means of either if employed alone.

Apart, however, from such general considerations there are conditions, some of which it would be difficult accurately to define, which determine the tolerance of either method or of both, and which must be taken into consideration by the medical attendant in order that the greatest attainable benefit may be secured for the patient. Physiologico-pathological conditions may give valuable indications. For example, distension and labour of the right heart dictate precaution with regard to such measures as are calculated to accelerate the return of the venous blood. The laboured breathing and sense of constriction in the direction of the epigastrium, which commonly follow the early immersions, and especially the first, are probably, in some degree at least, associated with the effort which is being made to restore due correlation of the pulmonic and systemic circulations. It will therefore be readily understood that, wherever an impediment lies between the two circulations, care and judgment must be exercised; and it is hardly necessary to observe that such a condition exists in cases of bronchial catarrh, asthma, emphysema, and pulmonary consolidation, whatever its cause, as well as where changes have affected the competence or sufficiency of the tricuspid and pulmonic valvular structures. Moreover, the impediment may act backwards, from the left heart, more especially in defect of the aortic valve mechanism. Again, rigidity of the coronary vessels resisting the impulse of an increasing systolic force and of a general stimulus to arterial expansion, may give rise to distress of the anginous order. The exceptional conditions of the cerebral vessels, both arterial and venous, in relation to contractility and adaptability to variations of pressure and volume, constitute a factor which must be taken into consideration with respect to functional disturbance and, in cases of deterioration of structure, to temporary congestion.

They also have an important bearing on cerebral anæmia and reparative nutrition, in relation to the power of sustained mental concentration, excitement, depression, and mental conditions generally,* as well as to the repair of lesions. It must not, however, be assumed that such conditions as have been enumerated contra-indicate treatment by the methods under consideration. On the contrary, they invite its application, but only with such careful adaptation and modification as may be indicated in each individual case.

It remains to take into account the phenomena of physiological reaction, due regard to which is one of the fundamental conditions of successful treatment. The temperatures which have been indicated as suitable for the baths are all below the normal of the human body. It will therefore be understood that it is not the object, and cannot be the effect, of the immersion to exercise an influence on the circulation by the exposure of the surface to a heat-communicating medium. On the contrary, it is sought to induce and establish therapeutic changes by a course which is precisely the opposite. Indeed it may be said that there is a difference of degree only, and not of kind, between the plunge which a hardy bather takes into freezing water and the careful immersion of a patient in a bath of a temperature a few degrees below that of the body. The question is: In what degree will a subject presenting, among other indications, certain morbid conditions, exhibit the physiological reactions which determine whether exposure to a cool medium shall be a source

* In this connection the reader may be referred to communications made to the *Lancet* by Dr. Maurice Craig and the writer, under the heading of "Blood Pressure in the Insane." June 25th and July 16th, 1898.

of injury or of well-being? The reply depends largely on the reserve of power which may at the time be latent in the sanguiferous organs and in the nervous system. Of the two the latter is probably the more important as regards the immediate result; while the state of the former, more especially with reference to tissue change, which may be irreparable, will largely determine the ultimate issue.

It has been suggested in the course of discussions which have from time to time arisen on this subject that the mineral and gaseous constituents of waters used for balneological purposes are inert as regards the human organism, and can in no way affect the result of immersion. That proposition is so distinctly traversed by balneological experts, as well as by universal experience, more especially as to the difference between fresh and sea water bathing, that it may be dismissed without further comment. The rubifacience they induce bears testimony to the influence which carbonic effervescing waters are capable of exercising on superficial vessels and nerves, and it can scarcely be doubted that such mineral constituents as the chlorides of sodium and calcium produce, to a minor and less apparent extent, a similar cutaneous excitation.*

A careful estimate of the patient's power of reaction, made with due allowance for such modifying local conditions as may exist, will determine, among other things, the temperature and mineral strength of the bath, the duration of the immersion and of the period of rest which must follow it, the number of baths which are to be taken in immediate succes-

* Since the above was written, Dr. Wilfrid Edgecombe and Dr. William Bain have published observations on the subject. (*Lancet*, June 10th, 1899, p. 1554.)

sion, as well as, later, the shortening or prolongation of the course, and the conditions under which the "after-cure" should be taken. The same consideration will, in the administration of the remedial exercises, determine the measure of resistance to be offered, the number of movements to be administered, the intervals to be interposed between each, the posture in which they should be executed, and whether one or more should be omitted in deference to such local conditions as renal or uterine displacement and ovarian congestion. It is obvious that the height to which the upper extremities should be raised must be made to bear some relation to the ability of the right heart to receive blood returning with increased volume and pressure from the uplifted arms.

In order to guard against injury and to ensure success, it is, therefore, necessary that the mind of the medical attendant should be specially directed to the phenomena of action and reaction. The effects of the immersion are to produce contraction of the cutaneous vessels and, in some cases, a consequent sense of cold, followed, after an interval which varies in different individuals and classes of cases, by dilatation of those vessels, an afflux of arterial blood, a glow of warmth, and a rise of general temperature of from one-half to one degree Fahrenheit. But it has to be borne in mind that the due sequence of reaction on action demands the exercise of a certain office on the part of the nervous system which can only be performed at the expense of a measure of energy already in store. The importance of these considerations will be appreciated when it is remembered that a large proportion of the subjects of chronic affections of the heart are also in greater or less degree neurasthenic. It may also be suggested that the stimulus which the therapeutic measures in question are calculated to convey to the heart must be considered in relation to the peripheral vessels, which in so many cases are the seat of degenerative changes involving a loss of conductivity, which may or may not in the end be susceptible to a process of repair, but which for the time being render them incapable of transmitting an increased supply of blood, and may, therefore, contribute to such a rise of intra-arterial pressure as would be a source of discomfort or even of danger to the patient. A sudden increase of arterio-capillary capacity, on the other hand, may be a cause of embarrassment to a myocardium which is prevented by structural changes, whether intrinsic or extrinsic, from making an adequate systolic response. In the former case, flushing, headache, excitement, or insomnia may ensue; in the latter, a sense of faintness or actual syncope. It cannot, however, be too emphatically stated that the avoidance of such drawbacks is within the capacity of the physician who is experienced in the practice of these methods, and that it should be his constant preoccupation to forestall their occurrence.

A good reaction is followed not only by a glow of warmth, but also by a sense of general comfort and of mental composure, which a period of rest in the recumbent position of not less than an hour's duration should increase and confirm. Indeed it may be desirable to prolong the period of recumbent rest. The experiments of Dr. George Oliver on man, confirmed by those of Mr. Leonard Hill on animals, show that, in the asthenic condition, radial enlargement is favoured by recumbency, as a consequence of the blood not being retained in the legs and the splanchnic area by gravitation. The importance of the part which the maintenance of a good reaction and of the uniform distribution of the blood may play in relieving an incompetent or dilated heart, is manifest.

Frequently the reaction glow takes place within two or three minutes of immersion. It may not occur until the patient lies down to rest. If it should be delayed beyond a few minutes from that time, it may be desirable to promote it by placing warm bottles near to the extremities, or one or more blankets over the body, and perhaps also by the administration of some warm stimulating fluid. Should such measures not be attended with success, the baths immediately succeeding should be modified as already suggested, and at the same time it would be advisable to inquire into the patient's manner of life in so far as it might affect the expenditure of nervous and muscular energy. Among definite symptoms of reaction-failure may be enumerated: a sense of cold throughout the immersion and persisting after it, yawning, and the occurrence of nervous rigor, such as is occasionally observed in and after parturition. They are important indications to the medical attendant of the direction in which the treatment should be modified, but they need by no means be accepted as proof that it is inapplicable. Indeed, reaction is a function which can be cultivated into steady and vigorous growth, and become confirmed into an endowment of infinite value to the patient, especially if he be gouty, rheumatic, or subject to frequent chills.

Enough has been said to show that the practice of these methods should not be lightly undertaken. Their safe and beneficial administration requires a careful adjustment of means to an end, and close attention to matters of detail. The novice who sets himself to test their value and to pronounce judgment on their efficacy, by experimenting on a few cases, and lighting on such as have been referred to in this chapter, is likely to encounter something more than discouragement, and to come to conclusions as unfavourable as they would be unfounded. It is obvious that to commit such treatment to nurses, however well trained, apart from the exercise of efficient medical direction and supervision, would not be in accordance with recognised medical ethics on the subject of the employment of unqualified assistance. That it should be carried out by unqualified persons on their own responsibility would be a grave abuse.

CHAPTER V.

CONDITIONS TO WHICH THE METHODS ARE APPLICABLE.

THE value of therapeutic measures which are capable of promptly relieving an over-burdened and labouring heart, without recourse to either general bleeding or dependence on drugs the use of which may sooner or later be attended with toxic or, to say the least, undesired effects, is so obvious that it does not stand in need of elaboration. Once it is recognised that the circulation, arterial, capillary, and venous, systemic and pulmonary, may thereby be stimulated to healthy activity and normal function, and that such health-restoring effects may, under careful and judicious direction, be maintained, progressively increased, and eventually confirmed without injury or drawback to the patient, it must be apparent that they may be applied with advantage to the greater proportion of the affections of the circulatory system, as well as to the relief of troubles which are not commonly considered to be exclusively cardiac in their origin and incidence.

Prof. Schott affirms that benefit may be expected to accrue in all cases of chronic heart disease, whether of valvular or parietal incidence, except where the myocardium has reached an advanced state of degeneration, or the vessels are the seat of advanced arterio-sclerosis. I have myself been witness of improvement amounting to practical or actual cure in cases presenting the physical signs usually regarded as indicative of the following affections: stenosis of either the aortic or the mitral orifice; stenosis of both; incompetence of either or both, with attendant dilatation; dilatation consequent on myocarditis, on habitual hæmorrhage and on constitutional anæmia; fatty heart; weakened heart; congenital mitral insufficiency; patent foramen ovale; and angina pectoris of apparently both neurotic and organic causation. It is reasonable to assume that measures calculated to diminish peripheral resistance, and to promote the nutrition and repair of the cardiovascular tissues, must be applicable to, at least, the early stages of aneurysm of the heart and great vessels.

The diagnostic and prognostic value of the exercises must not be overlooked. Familiarity with the effects which they may be expected to produce on healthy and on weakened walls enables the physician to detect early stages of dilatation, the existence of which it might otherwise be difficult or impossible to recognise. More pronounced dilatation may be readily differentiated from parietal hypertrophy, superincumbent fat, and pericardial effusion or infiltration. The measure of contraction induced by a few exercises readily discloses whether an abnormal area of dulness is to be attributed to dilatation or to a substantial mass of unyielding tissue. As regards prognosis, valuable information may be derived from the rate at which præcordial dulness is reduced, and, after a few days, by ascertaining, before the bath or exercises, as the case may be, the amount of more than temporary contraction which has been secured. Lastly, an unsuspected valvular lesion may be betrayed by the development of a bruit while the movements are in progress.

A question of no secondary importance is: May such recovery of heart-power and efficiency, together with the improvement in the general health which is contemporaneously effected by the systems under consideration, be so enduring as to justify, in the greater number of instances, the return of the patient to the cares and labours of an active life? From my own observation, and from the testimony of other observers, I am able to reply that such is the case.

In confirmation of these propositions it may be useful to adduce the evidence of cases illustrative of the effects of treatment in different descriptions of cardio-vascular affection. In so doing it should be premised that, whatever may be the histological condition of what is sometimes denominated "weakened heart," it cannot long persist without involving at least some appreciable degree of dilatation, dependent, probably, on softening of structure and loss of resilience. The cases are arranged according to physical signs rather than pathological conditions, which, in some cases at least, must in the present state of our knowledge, be matters of inference rather than of certainty.

CASE 1. *Weakened Heart.*—A lady aged fifty-seven had had fifteen accesses of influenza in about five years, three of which had occurred within the last twelve months. She complained of dyspnœa on exertion and on assuming the recumbent position, which, once set up, often persisted throughout the day or night. The area of dulness was increased to the right and left, and a feeble apex-beat could

CHRONIC DISEASES OF THE HEART. 49

CASE 1.

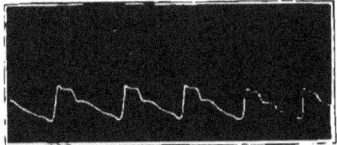

Before treatment. Hg. 168 mm.*

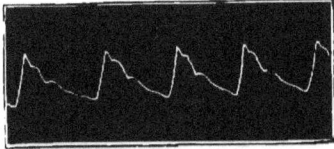

After 4th bath. Hg. 140 mm.

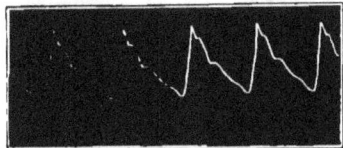

After 18th bath. Hg. 144 mm.

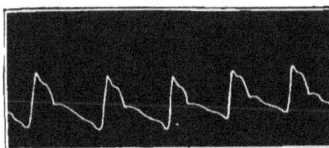

After 28th bath. Hg. 150 mm.

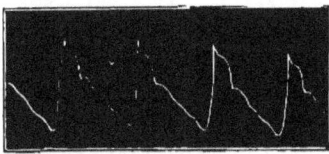

One month after treatment. Hg. 160 mm.

* The pressure, in this and the following cases, was taken with Hill & Barnard's larger sphygmometer.

be felt in the anterior axillary line. The heart-sounds were fœtal in character, and a systolic "whiff" was audible at the apex. She had twenty-eight baths, and her weakness was so great that, during the greater part of the course, she kept her bed in the intervals between the baths, which were mostly taken on alternate days. The pulse pressure, which at the commencement averaged 170 mm. Hg., came down to about 140. The area of dulness and the position of the apex-beat returned to normal, and, after a stay of nearly three weeks at Crowborough, the patient returned to London in fairly good health, and was able, with care not to overtax her strength, to lead an ordinary life. In this case the chest was marked with cutaneous angiomatous macules, the significance of which has been alluded to elsewhere.*

CASE 2. *Angina sine Dolore.*—A country clergyman, an old sportsman, seen in October, 1896, then fifty-six years of age. He could not walk more than from 150 to 200 yards without being arrested by præcordial distress and a sense of impending death. He had had several accesses of syncope, and, three weeks before coming to London, had fallen unconscious after dinner while being assisted upstairs by two of his sons to his bedroom on the first floor. On reaching the room he fainted again. The bowels acted once daily, the stools being consistent and yellow. He suffered from a gnawing sense of hunger before meals. On his coming to London I visited him with one of my bath assistants, and endeavoured to administer two of the arm movements with the lightest resistance; while he

* "Cardio-vascular Repair." J. & A. Churchill.

CASE 2.

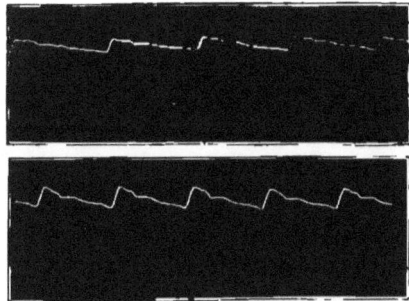

Before and after exercises (2nd day).

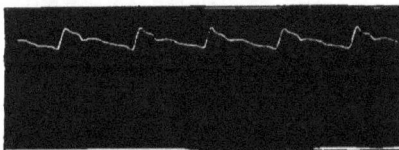

Before and after 10th bath.

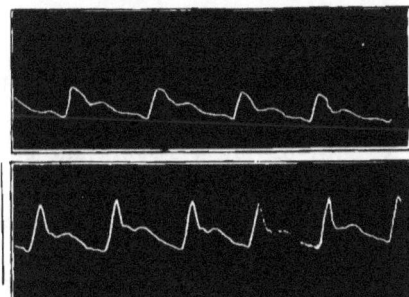

Before and after 19th bath.

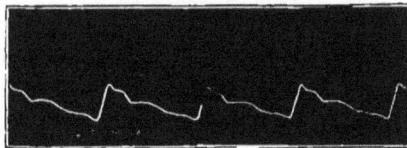

One hour after last bath.

was executing the second he fell into our arms, and we laid him unconscious on the sofa. As soon as he regained consciousness we administered slight movements of the hands and fingers, while he remained in the recumbent position. These and others I ordered to be repeated for three days still in the same position. On the fourth day I was present when he had his first mineral bath; it was followed by a good reaction. In three weeks' time he could not be restrained from the enjoyment of walking three and four miles daily. After a course of twenty-eight baths he returned home, and about Christmas time started on a visit to friends in India. In June, 1897, a mutual friend informed me that he was in perfect health, and on the Saturday preceding he had made ninety-eight at the village cricket-match, and on the following day had taken four services single-handed. I heard of him again in June, 1899. He was in excellent health, and fulfilling all his clerical duties, as well as leading the active life of a country gentleman.*

CASE 3. *Angina cum Dolore.*—F., æt. sixty-three, proprietor of a business in the West-end of London, first seen on the 13th of May, 1898, had suffered for more than five months from acute pain, extending from the præcordium to the left arm, which supervened on all but the smallest exertions. He could not carve at table without bringing it on. Most mornings he had been obliged to stand still for some minutes in the course of a walk of a few hundred yards taken to catch an omnibus, and had been able to proceed at a slow pace only after inhaling amyl nitrite and swallowing a nitro-glycerine

* This is Case III. reported in " Cardio-vascular Repair."

CHRONIC DISEASES OF THE HEART.

CASE 3.

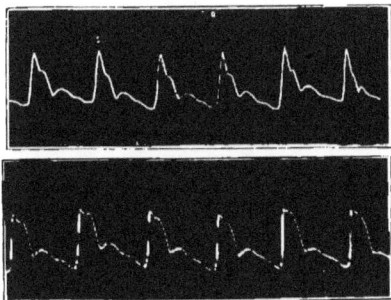

Before and after 1st bath.

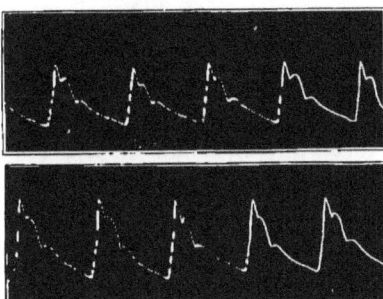

Before and after 7th bath.

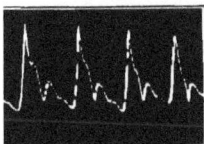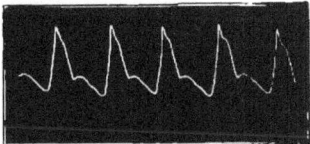

Before and after exercises, ten days later,
with acute coryza (influenza?).

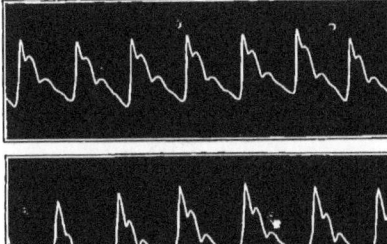

Before and after the last (28th) bath.

tabloid. He had been under medical treatment, and advised to dispose of his business. After nine days of preliminary treatment, in the course of which the pulse-pressure fell from 170 mm. Hg. to 130, he commenced a course of baths and exercises. On the 27th he reported that he had taken neither tabloid nor inhalation for seven days. Two or three times during the remainder of the treatment recourse was had to those remedies. On July 21st he presented himself, after a stay in the neighbourhood of Malvern, and stated that, a few days after his arrival there, he had walked without discomfort to the top of the Worcestershire Beacon and back, that he had resumed his usual business occupations, and only on rare occasions experienced any reminder of the old sensations. In the January following I saw him two or three times with Dr. Wightwick, in the course of a severe attack of influenza. He made a good recovery, and has since reported himself as being "very well." Throughout the treatment, and on his return from the country, the pulse-pressure was found to range from 130 to 136 mm. Hg.

CASE 4. *Angina cum Dolore.*—A man, aged fifty-eight, was examined in September, 1898, by Dr. Lauder Brunton, who favours me with permission to quote as follows from his notes, which were headed "Atheroma:"—"Complains of albumen in urine, from a trace to a cloud. For twenty years has been subject to 'gout.' Acute nephritis fifteen years ago. Attacks, more like rheumatism than gout, lasting three or four months at a time, and presenting no redness and but little swelling. Apex half an inch outside the nipple-line. Loud systolic murmur, subsiding after recumbent position has been assumed. Urine pale, clear, acid, 1013, thin cloud of albumen."

On the 4th of February, 1899, I found the apex-beat half an inch outside the nipple-line. There was no bruit, the pulse was 108 (sitting position), pressure 210 mm. Hg. No albumen. He reported—"August 11th last, rose early to catch a train, having sat up late the previous night. While driving to

CASE 4.

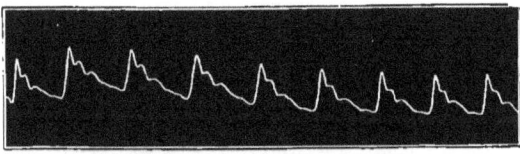

February 4th.

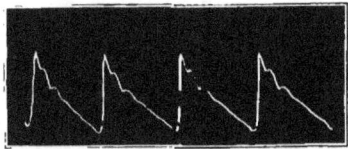

March 17th.

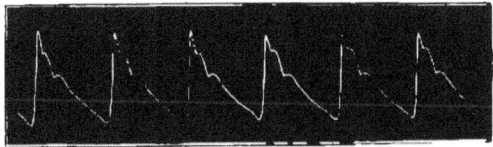

April 5th.

the station, a burning pain came on in the region of the heart, and remained very severe for five or six minutes. At the end of the journey, under the influence of some annoyance, the pain became very severe for about ten minutes. The next day, went out after luncheon and shot twelve brace of grouse; but, after walking about fifty yards up a hill, the pain came on again very severely for ten minutes. Ever since has been subject to a 'catching' feeling and some burning whenever exertion is made, and cannot walk a hundred yards without bringing it on."

In conjunction with a course of baths and exercises, he took small doses of iodides, with sodium bicarbonate and liquor bismuthi, and for six consecutive nights gr. ⅛ of calomel. By the 17th of March he could walk for one and a half hours without discomfort. He was last seen on the 5th of April, after a fortnight's after-cure at West Malvern, and reported that he had been able to walk to the top of the Worcestershire Beacon without pain or fatigue, but that on one or two occasions he had experienced a suspicion of the old sensations on attempting to take a brisk walk immediately after breakfast.

In the following tables that referring to the frequency of the pulse gives, in the first column, the frequency in the sitting posture, and the second and third the frequency as counted in the quarters of a minute immediately following the act of standing up.

		Frequency.			Pressure.		Body-weight.
		1st ¼ m.	2nd ¼ m.	3rd ¼ m.			st. lbs.
Feb.	4.	27	27	27	210 mm. Hg.		13 13
,,	16.	16	19	19	170	,, ,,	13 10
,,	27.	18	19	18	160	,, ,,	13 8½
March	10.	18	19	19	170	,, ,,	13 6¾
,,	17.	16	18	17	170	,, ,,	13 6¼
April	5.	16	18	17	160	,, ,,	13 4½

CASE 5. *Apex-Systolic Bruit in Mitral Area.*— A man, aged forty (patient of Dr. Lucas, Bury St. Edmunds), had had acute rheumatism in childhood, and twice in manhood. Dr. Lucas had attended for endocardial mischief. For the last six months has been rapidly declining in health and strength,

CHRONIC DISEASES OF THE HEART

CASE 5.

Before.

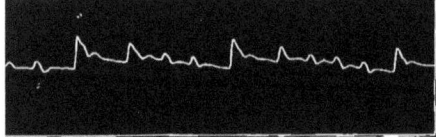

After,

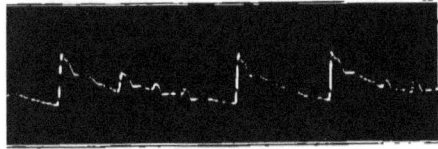

1st bath, October 22nd, 1897.
Before.

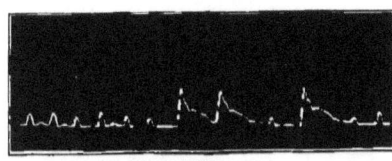

After,

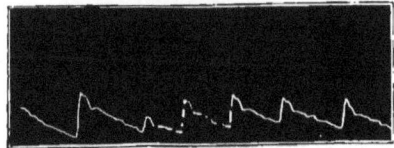

3rd bath, October 25th.
Before,

CASE 5.—*continued.*

Before.

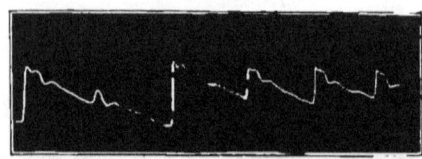

After,

9th bath, November 2nd.

Before.

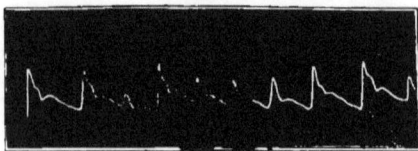

After.

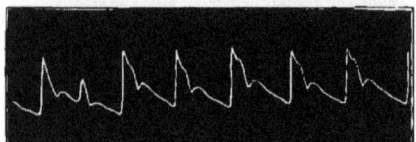

15th bath, November 15th.

Before.

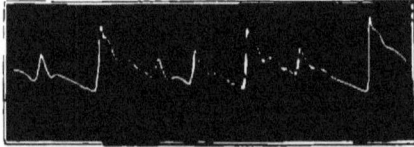

After,

28th bath, November 28th.

with increasing dyspnœa, which now supervenes on the slightest exertion. He entered the room with difficulty, supported by a companion, and with severe dyspnœa and a dusky face. On the 27th of November, 1897, he commenced a course of baths and exercises, on the conclusion of which, on the 11th of November, he went to West Malvern. Shortly before he was to have left that place he was taken with acute bronchitis, and, as soon as he was able to travel, returned to Bury St. Edmunds. Before the course of treatment the apex-beat was $1\frac{3}{4}$ in. outside the nipple-line, and at the conclusion, $\frac{1}{4}$ in. On the 15th of June in the following year, I was favoured by Dr. Lucas with the following report:—" When —— returned from Malvern he was suffering considerably from dyspnœa, and his heart appeared to be more irregular. I kept him in bed for a month, and since then he has been better all round. His cardiac dulness is the same as when you last examined him. His heart has undoubtedly maintained the improvement which it had undergone from the treatment. He is able to attend to his business, under restrictions. A few weeks ago he underwent one of the severest conceivable of mental shocks. I quite expected that it would have had a very bad effect, but up to the present he does not seem to be any the worse."

CASE 6. *Traumatic Lesion of Aortic Valves.*—An omnibus driver, aged forty-eight, reported, on the 10th of February, 1899, that until six weeks previously he had been in good health, when, on leaving the box after making a great effort to hold up both of his horses, which had slipped at the same moment, he found that he could scarcely walk for shortness of breath. He had remained under medical treatment in the interval, could scarcely walk ten paces, and spent a great part of each night in a condition of orthopnœa. He commenced a course of baths and exercises on the 13th. By the 18th he began to enjoy unbroken nights, had recovered his appetite, and could walk better. On the 23rd of March he completed his treatment, and went into the country in good general health, and able to walk a quarter of a mile at a good pace without discomfort. He returned on the 25th of April somewhat further improved, to apply for the position of a timekeeper in the service of his company. At the commencement of the treatment there was audible, at a distance of from six to eight inches from the chest, a loud musical murmur, accompanied by a thrill, which was to be felt over the greater part of the front of the chest wall. On his return it was barely audible at a distance of two inches. It was diastolic in time, and preceded, without appreciable interval, by a slight sh-sound.

CASE 7. *Double Aortic Bruit with Mitral Systolic.*— This case was reported in the *Lancet* of March, 1894, as follows, and has been under observation up to the present time (May, 1899), when the patient has just completed her seventeenth course of treatment:—" A woman, fifty-two years of age at the present time, and the subject of inveterate lithæmic tendencies, rapidly developed, in the winter of 1891-92 a loud, rasping, basic systolic bruit, which was accompanied by a systolic apex souffle. Her health rapidly failed, and the cardiac condition, involving as it did loss of sleep and appetite and steadily increasing dyspnœa, threatened to bring her life to a close. All ordinary resources having failed to afford relief, I suggested recourse to the Nauheim baths and treatment by

CHRONIC DISEASES OF THE HEART. 61

CASE 6.

Before.

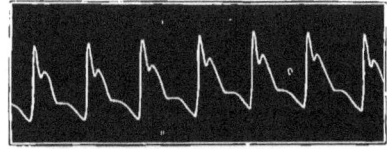

After.

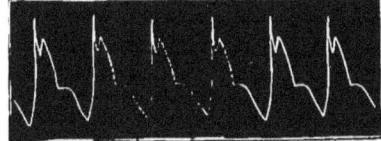

1st bath.

Before.

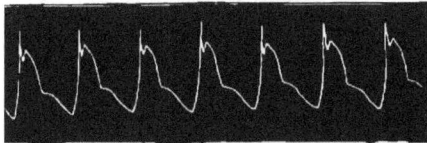

After,

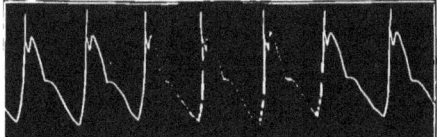

7th bath.

Before.

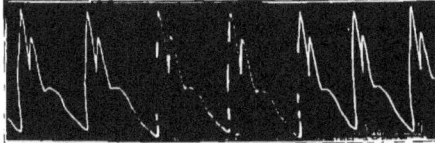

After,

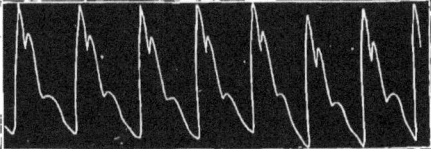

28th bath.

·exercises, with the practical details of which I was then unacquainted. The journey, however, appeared to involve such serious risk that I did not venture to authorise it; but, as Dr. Hermann Weber came to the conclusion that it probably afforded the only remaining chance of life, it was decided to undertake it. When again seen in the following October, the patient was restored to her former measure of health and pursuing her usual avocations. In May, 1893,

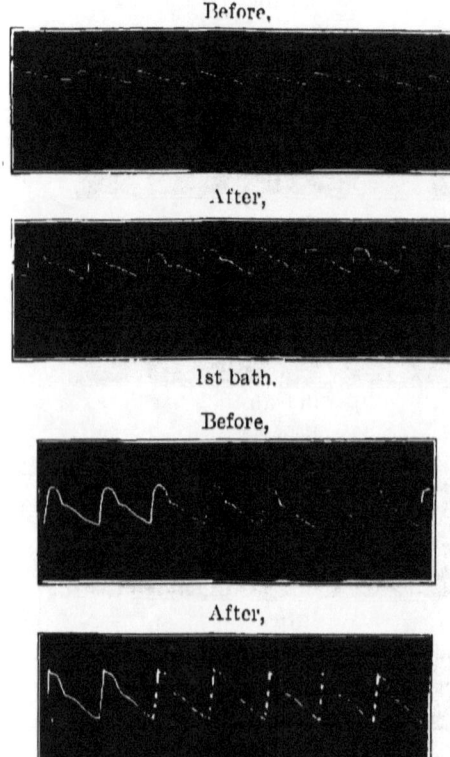

CASE 7.

Course V. December, 1894—February, 1895.

Before,

After,

1st bath.

Before,

After,

28th bath.

CHRONIC DISEASES OF THE HEART. 63

CASE 7—continued.
Course VI. March to May, 1895.
Before,

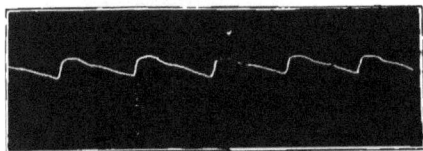

After,

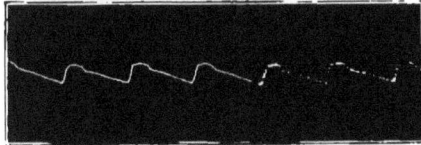

1st bath.
Before,

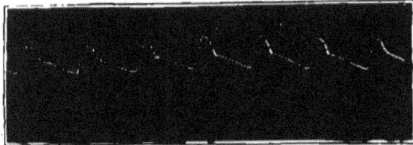

After.

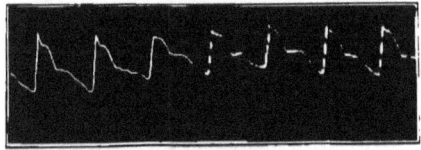

23rd bath.
Course X. October to November, 1896.
Before,

CASE 7, Course X.—*continued*.

Before.

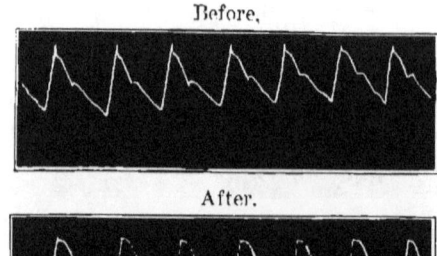

After.

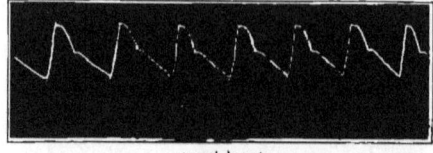

22nd bath.

Course XIII. November, 1897, to January, 1898.

Before,

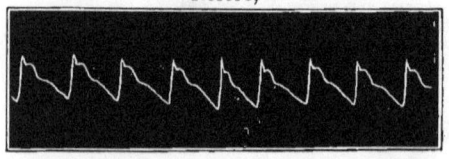

After,

1st bath.
Before,

CASE 7—continued.
Course XVI. October to December, 1898.

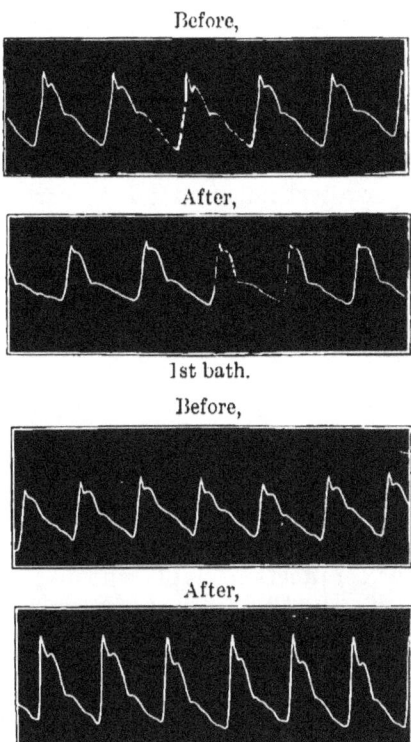

Before,

After,

1st bath.
Before,

After,

28th bath.

she returned to Nauheim for a second course, and on her return I was unable to detect either the basic or apex bruit, and the heart was fully competent. In the middle of November a severe access of influenza, unaccompanied by catarrh, and the whole incidence of which appeared to be on the heart and nerve centres, was followed by a return of the bruits and incompetence." A course of exercises was followed by the establishment of a fair measure of compensation, with cardiac competence and a return to rather more than the former measure of health and activity. A further access of influenza in 1894 necessitated another course of treatment at Nauheim in the summer of that year. From that time the heart has never suffered dilatation or incompetence ; but the lithæmic tendency has never been overcome, and, two or three times a year, has culminated in such severe articular pain, with periosteal inflammation of portions of the cranial bones, the clavicles, the iliac crests, and sacro-iliac synchondroses, that, other remedies failing, the patient has sought and found relief in successive courses of baths and exercises, at one time in Nauheim and at another in London, according to convenience and the season of the year. From the time of their recurrence, above reported, the murmurs have persisted and scarcely varied in intensity.

CASE 8. *Apex Systolic Bruit extending to the Left Sternal Margin.*—A patient of Dr. J. Lumsden Propert (in conjunction with whom each of the following observations was made), twenty-seven years of age, was a tall man with a chest proportionately somewhat narrow, who found himself to be debarred from active exertion by dyspnœa and palpitation. There was a well-marked upheaval of the chest-wall over the right ventricle, with epigastric pulsation. A systolic bruit, loudest about midway, was audible from the apex to the sternal margin. The first sound at the base was inaudible. After the fourth bath the bruit was found to be diminished in intensity, but there was marked reduplication. The first sound to the left of the apex-beat was clean. On the occasion of the tenth bath the first sound at the base had become audible. Fourteen days later (after twenty-second bath) neither bruit nor reduplication was audible at the apex.

Seven days later, the treatment having concluded, there was slight reduplication, but no bruit. The general health had improved, and the patient was able to take long walks without either dyspnœa or palpitation. The apex-beat, which before treatment had been in the sixth space and three-quarters of an inch within the nipple-line, was in the fifth space an inch and a quarter within. There remained neither upheaval nor epigastric pulsation. Dr. Propert reports that the improvement has been maintained.

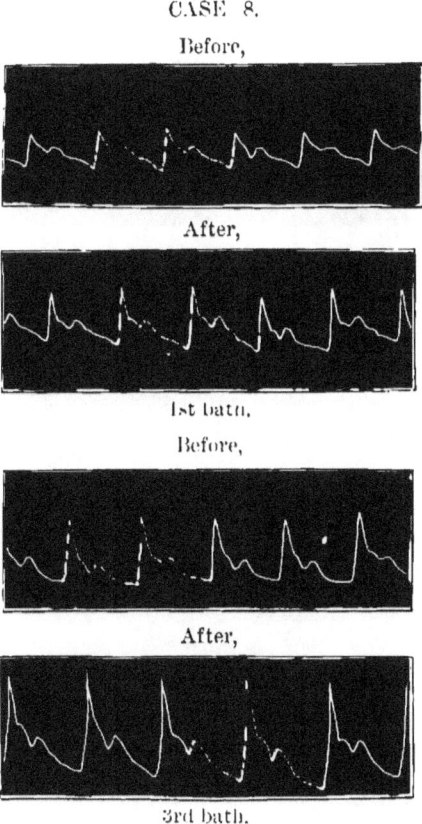

CASE 8.
Before,
After,
1st bath.
Before,
After,
3rd bath.

CASE 8—*continued.*

Before,

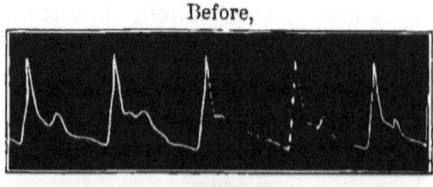

After,

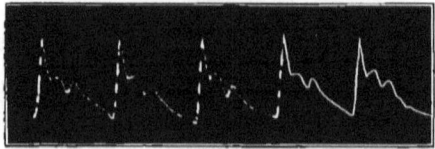

10th bath.

Before,

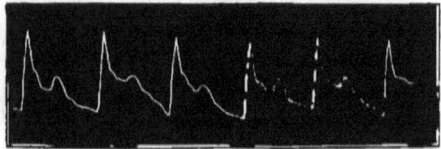

After,

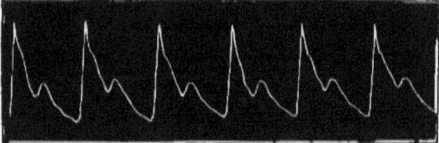

22nd bath.

Before,

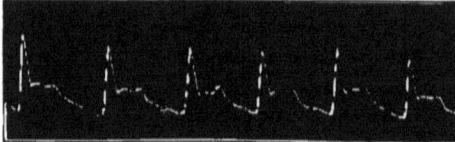

After,

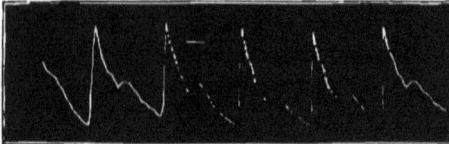

28th bath.

CASE 9. *Athlete's Heart.*—A young man, aged eighteen, with whom there had been difficulties of digestion in infancy and throughout childhood, but who was well grown, was pallid, and expressed himself as being unable to sustain any ordinary exertion

CASE 9.

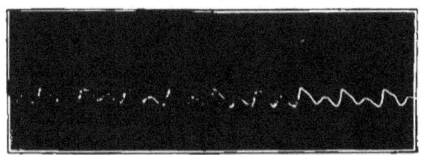

Before treatment, December 5th, 1898.

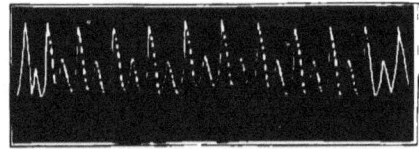

On conclusion of treatment, January 25th, 1899.

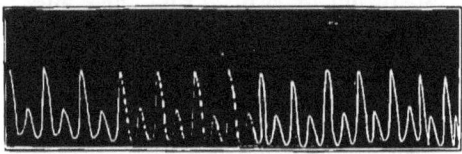

March 13th, 1899.

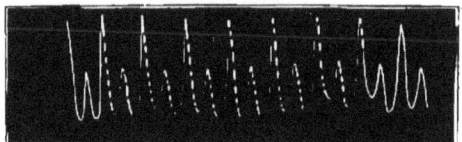

May 1st, 1899.

since, a year before, he had made a cycling tour, in the course of which he had frequently covered ninety miles a day, on irregularly taken and often insufficient food, and from four to five syphons of aërated water daily. He did not suffer from dyspnœa on exertion. The maximum apex-beat was one inch within the nipple line. There was upheaval of the whole left mammary region with each pulsation, the frequency of which at the heart was 160 and over, and at the wrist 120. The first sound at the apex was blunt. He derived marked benefit in health, energy, and appearance from a course of baths and exercises under Dr. Schott, at Nauheim. Three months later (December, 1898) he came to London for a further course, as there were signs of relapse. The pulse frequency then was 140 at the heart, and from 130 to 136 at the wrist. The first bath was taken on December 5th. After the second bath the frequency was the same at the heart and wrist—130. On the conclusion of the course it had fallen to from 110 to 120; there was no upheaval of the mammary region, but an exaggerated impulse was to be felt from the apex, one inch within the nipple line, to the sternal margin. The face and lips were ruddy and the general health good. When last seen, on the 1st of May, 1899, he appeared to enjoy perfect health and vigour, and was playing golf almost daily. The pulse remained at 120 at both heart and wrist. It is proposed that a further course of treatment shall be taken at Nauheim in the course of the summer. In similar cases it has been found that two or three years are required for complete recovery, and that there is danger of relapse during the period of growth and development. Throughout the course a systolic respiratory bruit, commencing, as regards distribution, one inch outside the nipple-line, and increasing in harshness and intensity to within one inch of the spine, and fading above and below the sixth interspace, was audible throughout the inspiratory act. It underwent a gradual diminution of intensity, and on May 1st could be heard only at the angle of the scapula, with the first systole occurring in each inspiration.*

CASE 10. *Myocardial Degeneration with Arterio-Capillary Sclerosis (?) and General Soakage.*—A lady, aged fifty-three, a patient of Dr. Forbes Fraser, of Tarporley, who reported that during the two years she had been under his care there had been occasional accesses of faintness and palpitation, sometimes associated with cardiac pain, culminating, in September, 1898, in a severe attack of angina following mental shock. He added that from that time the accesses had been more frequent and severe, and that a very grave attack had taken place in the following November. Dr. Fraser expressed the opinion that the heart was in a state of fatty degeneration. The patient's father had died of angina pectoris at the age of sixty-eight, her mother's sister and brother had died of sudden failure of the heart. She stated that her brother suffered from heart-pain. Her own account was to the effect that from the age of twenty-one she had been subject to neuralgia and to fronto-occipital headaches, that since a bad confinement, six years previously, her ankles had swollen in the evening, and that the last access of syncope had been accompanied by severe pain extending from the nape of the neck to the lowest part of the spine. There had been an attack of pericarditis in 1883. The face and hands were puffy, as were the legs and ankles. The puffiness was of a brawny character, and as a rule there was no pitting until the latter part of the day. The apex-beat was not perceptible in the prone position, but cardiac dulness did not extend beyond the nipple-line. The sounds were distant and of fœtal character, the second sound at the apex

* At the time of going to press the respiratory bruit had ceased to be audible.

was barely audible, the second at the base relatively accentuated. The bowels acted twice daily, the motions being yellow, and the second either fragmentary or liquid. The pulse was regular in frequency, ninety in the sitting posture, ninety-six on standing up, and fell to ninety-four within the minute ; the pressure was 210 mm. Hg. Treatment by baths and exercises commenced on November 15th, 1898, and throughout the course its effects were marred by a series of disturbing causes. I was dissatisfied with the result, and although Dr. Fraser was struck with the improvement which he observed on the patient's return, I strongly urged a second course, which should be conducted under conditions of isolation, to guard against all influences calculated to depress the nervous energies. The second course commenced on the 12th of January, 1899, and terminated on the 12th of February following. The patient then went to West Malvern for a stay of rather more than a fortnight. On the 24th she reported that, although, before treatment, she had only been able to crawl about a dozen yards on the level, she could then take the steepest

CASE 10.
First Course.
Before,

hill without discomfort. I saw her again on the 3rd of March. She was in good health, and the only remaining puffiness was to be found behind the ankles. There was no pitting. The day before leaving West Malvern the patient had walked six miles without discomfort or undue fatigue. The body-weight, which before treatment had been 11 st. 4½ lbs., was 10 st. 8 lbs. On May 13th, Dr. Fraser reported: "She remains, as far as I can see, perfectly well." *

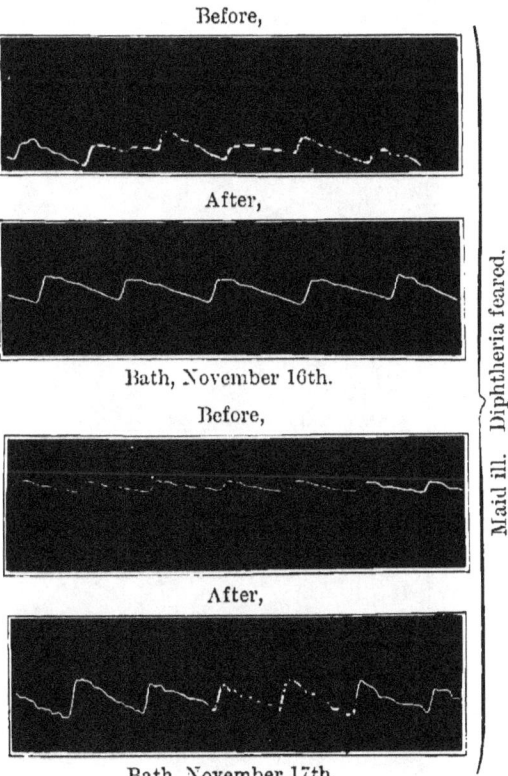

CASE 10, First Course—*continued.*

* In the course of the treatment the pulse-pressure fell to 180 mm. Hg., but after its conclusion rose again to 210 mm. Hg.

74   THE SCHOTT METHODS OF THE TREATMENT OF

CASE 10, First Course—*continued.*

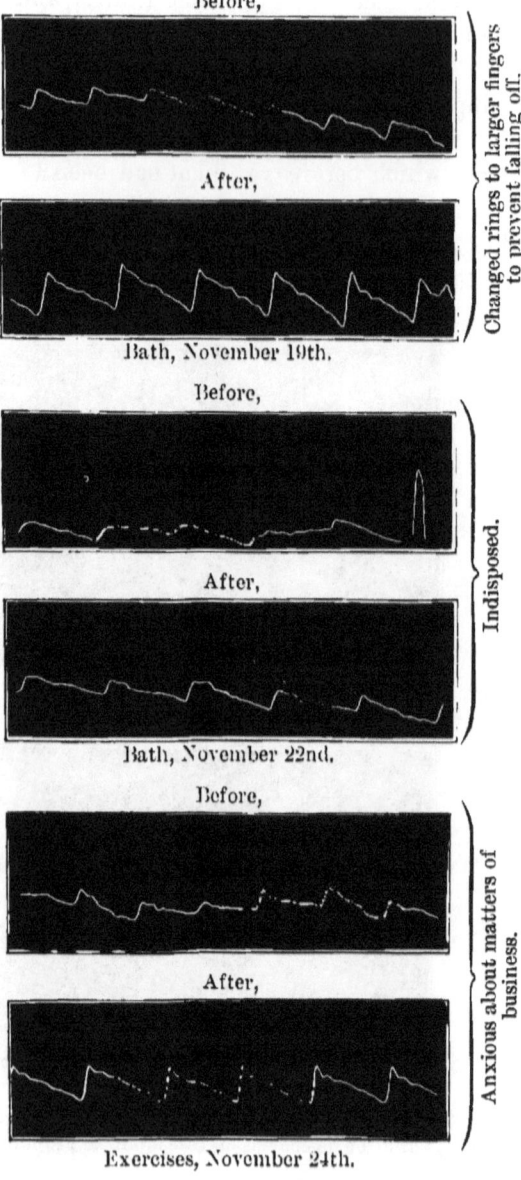

Bath, November 19th.

Bath, November 22nd.

Exercises, November 24th.

CASE 10, First Course—*continued.*

Before.

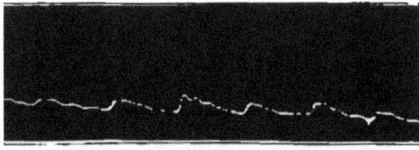

After,

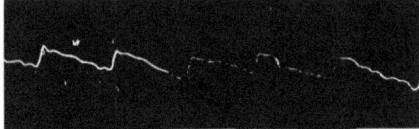

Bath, November 26th.

Before,

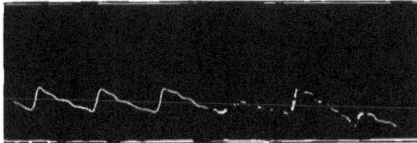

After,

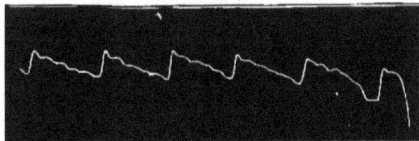

Bath, November 29th.

Before,

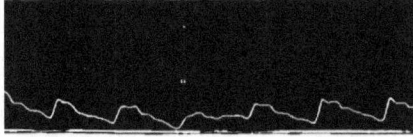

After.

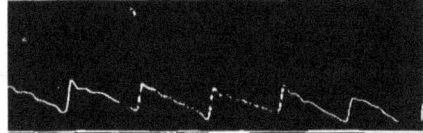

Exercises, December 1st.

CASE 10, First Course—*continued.*

Before,

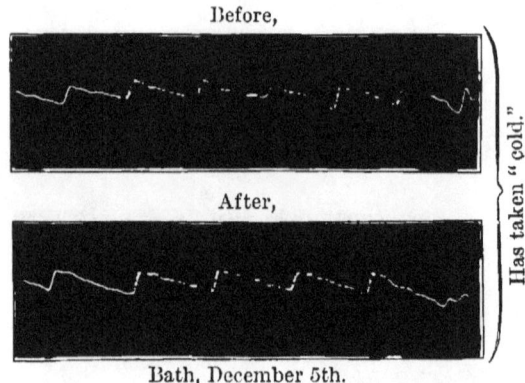

After,

Bath, December 5th.

Has taken "cold."

Before,

After,

Exercises, December 8th.

Before,

After.

Bath, December 10th.

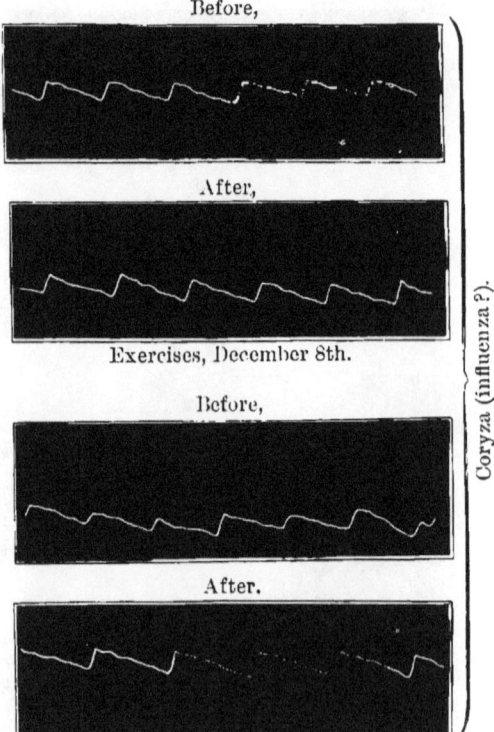

Coryza (influenza?).

CHRONIC DISEASES OF THE HEART.

CASE 10, First Course—*continued.*

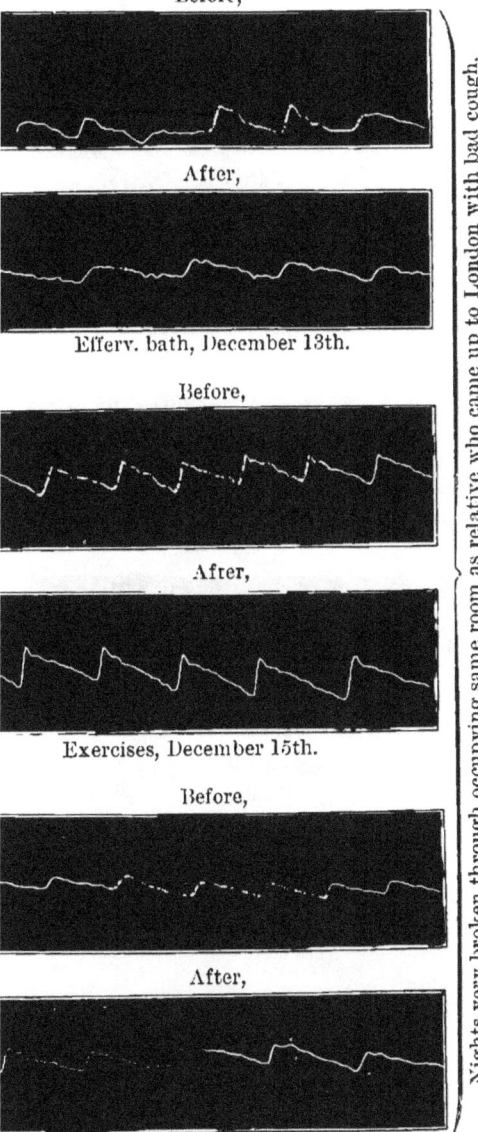

Before,

After,

Efferv. bath, December 13th.

Before,

After,

Exercises, December 15th.

Before,

After,

Efferv. bath, December 17th.

} Nights very broken through occupying same room as relative who came up to London with bad cough.

CASE 10, First Course—*continued*.

Before,

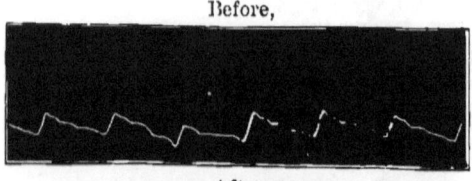

After,

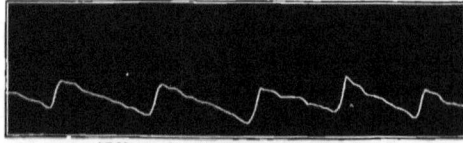

Efferv. bath, December 19th.

Before,

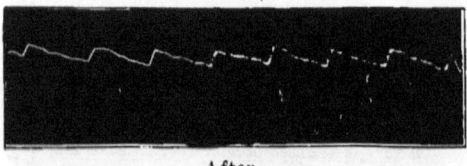

After,

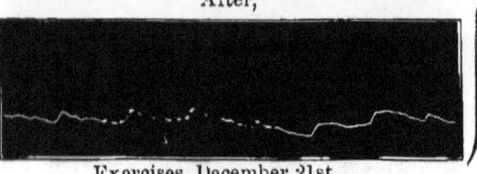

} After emotion on receiving disturbing letters.

Exercises, December 21st.

Second Course.
Before,

CASE 10, Second Course—*continued*.
Before,

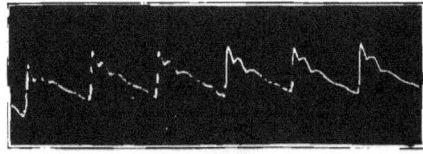

After,

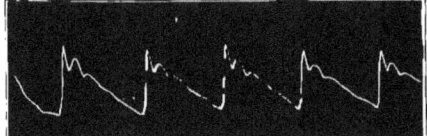

Bath, January 14th.
Before,

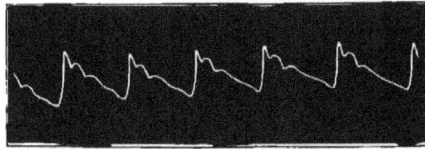

After,

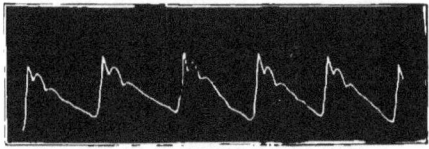

Bath, January 18th.
Before,

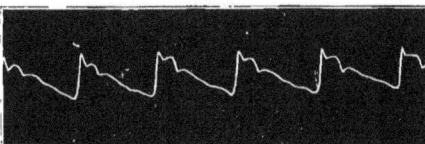

After,

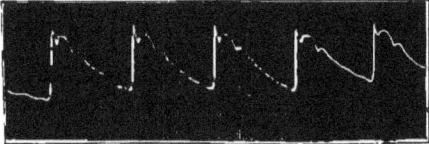

Efferv. bath, January 23rd.

CASE 10, Second Course—*continued.*

Before,

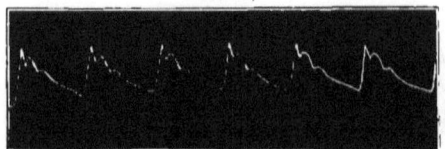

After,

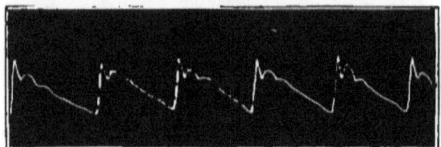

Efferv. bath, January 28th.

Before,

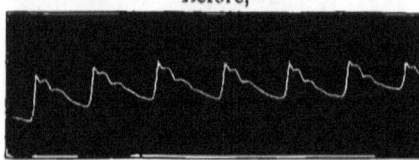

After,

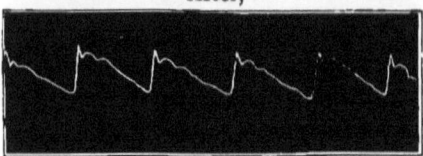

Efferv. bath, February 3rd.

Before,

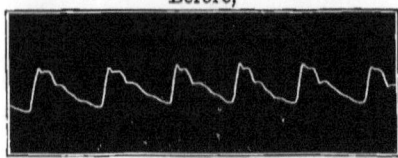

After,

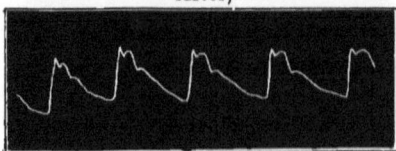

Efferv. bath, February 7th.

CASE 10, Second Course—*continued*.

Before,

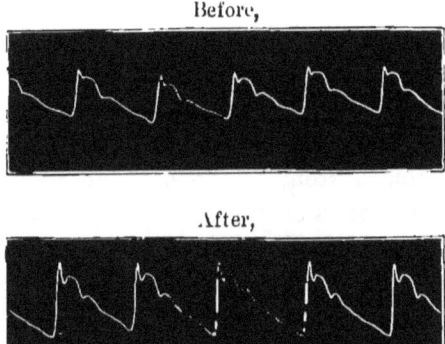

After,

Last bath (efferv.), February 12th.

The whole of the sphygmograms which were taken in this case are reproduced with a view to emphasising the importance of isolating patients, during their bathing course, from social, domestic, and business concerns. That end, it need scarcely be pointed out, will, in a great proportion of instances, be best attained by resort to a foreign watering-place, such as Nauheim. The pulse-tracings also serve the purpose of testifying to the efficiency of the Schott methods in promoting the repair of degenerated vascular as well as of cardiac tissues.

Such examples of the effects of the methods under consideration, in similar cases and in others, such as those of Graves' disease in early and in advanced stages, might be adduced almost indefinitely. To do so could only weary the reader. He is invited, however, to consider them, and especially the sphygmographic tracings, by means of which the course of recovery, and in some cases of relapse, is illustrated from a point of view which has not as yet been insisted on, namely that of vascular repair. As arterial lumen increases, whether in obedience to vasomotor influence or to systolic force, or to both combined, the vascular tunics undergo repair and rejuvenation. Nay more, patency may be restored to occluded arterioles and capillaries. The process may be watched from day to day and from week to week in the vessels of the face, more especially those of the nose and cheeks, and in the zones of varicose capillaries which often mark the supra- and infra-mammary regions in subjects of atheromatous and other forms of degeneration.* It is reasonable to assume that similar changes take place in the coronary and myocardial circulation, in the great nerve centres, in the lungs and other viscera, and throughout the system generally. The clinical evidence afforded by numerous cases of myocardial degeneration with and without angina, of paresis and paralysis, of asthma, emphysema, and pulmonary consolidation, of albuminuria, of retinal hæmorrhage, and of impairment of mental faculty, confirm the assumptions based on superficial and obvious anatomical changes. Among morbid conditions none yield more readily and completely than those vascular degenerations which have not proceeded to calcification or to irreparable fibrotic change.

* This subject is dealt with at greater length in "Cardio-vascular Repair." J. & A. Churchill.

CHAPTER VI.

CONDITIONS NOT PRIMARILY CARDIAC TO WHICH THE METHODS ARE APPLICABLE.

In view of what has been said in the preceding chapter it remains to enumerate some of the conditions, not primarily or ostensibly cardiac, to which the methods under consideration may be applied with advantage.

First in order, and perhaps most obviously, come those which are associated with the presence of uric acid in the blood in excess. Such are those in which the sequelæ of acute rheumatism, and especially peri-, myo-, and endocardial lesions co-exist with injury to joints and tendons. What has been said of the resolvent, metabolic, and trophic effects of the baths points to them as remedial and restorative agents of the first order of efficacy, especially when their direct effect in contracting the heart and nourishing its tissues is borne in mind. Where myocardial changes have left a legacy of dilatation and feeble action, and the structures entering into the composition of the joints have either escaped injury or enjoyed the advantages of effectual resolution, the exercises alone may meet the requirements of the case. The same indications apply to the subjects of acute and chronic gout. Cases of dilatation, weakened heart, or special lesion occurring in the course of osteo- or rheumatoid arthritis stand, however, on different ground. When the central and trophic influences which, according to Dr. W. M. Ord and others, play a prominent part in this affection, as well as the measure of neurasthenia and the profound cachexia which frequently arise in conjunction with the arthritic changes characteristic of the disease, are taken into consideration, it will be apparent that the treatment by baths is calculated to relieve, at one and the same time, such cardiac troubles as may have arisen, and the other local and general conditions to which allusion has been made.

The weakened heart which influenza so often leaves in its train, probably as the result of myocarditis and of altered innervation, is readily amenable to both baths and the exercises. In some such cases murmurs may be detected, mostly basic-systolic. They are probably due to orificial irregularity or dilatation, for they are generally subdued in the first stage of the treatment. Sometimes they recur in the course of the earlier intervals; but I have not met with any, not previously existent, which have survived a full course (Cases F and G, Ch. VIII.).

The effects of both baths and exercises, whether singly or conjointly employed, are very remarkable in anæmia associated with more or less dilatation, whether chlorotic, gouty, malarial (Case C, Ch. VIII.), or arising from loss of blood or from chronic intestinal catarrh. In many such cases it is common to see the colour, digestion, spirits, energy, and general health of the patient undergo a notable improvement within three or four days, without the exhibition of either arsenic or iron. A course of four or five weeks, combined with due precaution as to diet, exercise, and general hygiene, is usually sufficient to ensure a return to health. In the two latter classes of cases, it need not be said that the physical treatment should be combined with measures calculated to arrest the waste which the system has suffered. The effects of both methods, especially when combined, have been no less satisfactory in such cases of œdema, anasarca, serous effusion, and albuminuria associated with deficient heart power as, under my observation, have submitted to the treatment. Case M. (Ch. VIII.) is one in point.

The Schott methods have brought relief to such cases of asthma, associated with however little cardiac dilatation, as have come under my treatment. I may mention three typical cases; a is a lady of middle age who, on taking a drive, or in any way coming near a horse, experienced the following train of symptoms: — intense injection of the ocular and palpebral conjunctivæ, nasal defluxion, hoarseness, and the breathing characteristic of spasmodic asthma. After a week of baths she was able to take a long drive with relative impunity, and, as the course proceeded, the improvement continued, until finally the symptoms were scarcely appreciable. A slightly dilated heart had resumed its normal dimensions within the first week, and the pulse had become uniformly stronger and fuller.* β is a lady, thirty-six years of age, who has been liable, with increasing frequency and severity, to accesses of eczema, intestinal catarrh with abilious stools, bronchitis with profuse muco-purulent expectoration, and asthma with nocturnal exacerbations of great severity. The effects of driving were similar to those experienced by a. When she came under treatment she had not been able to lie down for a fortnight, and could secure only a few snatches of sleep with the aid of the

* The benefit has, in this case, been maintained for four years.

fumes of a well-known anti-asthmatic powder. The lungs were emphysematous, and the apex beat was two inches outside the nipple line. During the day, any slight exertion induced cardiac dyspnœa and præcordial distress. Within three days of commencing a course of exercises, all the symptoms had so improved that she could lie down at night and obtain unbroken sleep for two or three hours at a time. At the end of a fortnight the anti-asthmatic inhalations were discontinued. On the conclusion of a course of five weeks (inclusive of the menstrual interval) she enjoyed good nights, could go up and down stairs without the breathing being affected, take long walks, and drive, with scarcely appreciable inconvenience, through the streets of London in hot, dry, and dusty weather. The apex beat had receded two inches and was in the nipple line.* γ differs from the preceding in having passed the climacteric period by about two years, and in the emphysema being more pronounced. For seven years she had only obtained sleep by being pillowed up and inhaling the fumes of nitre-papers, and outdoor exercise had been limited to slow, rambling walks in the garden of her country residence. The exercises alone were employed. The nocturnal asthmatic exacerbation, from day to day, occurred a little later and lasted a shorter time. In the course of the second week, the nitre-papers were abandoned, and good nights were enjoyed in the recumbent position. On the conclusion of a five weeks' course, the apex beat had receded from an inch without to half an inch within the nipple line. When last heard of, three years later,

---

* This patient has enjoyed better health for three and a half years, but, after an attack of bronchitis, had a course of baths at home.

the patient was taking country walks of about a mile every day, and leading a fairly active life.*

I have mentioned these cases with some detail because they seem to open up a prospect of relief to a class of sufferers whose troubles have hitherto, to a great extent, defied treatment. The results recorded are, however, not surprising when considered in the light of what has been shown to take place in the relief of a burdened heart, and the improvement of the capillary circulation. It can scarcely be doubted that the congested and varicose veinlets which encumber the alveoli in such cases, share in the general change for the better, and that the circulation through the pulmonary circuit is quickened, and the aëration of the blood proportionately facilitated, by the increasing systemic arterial and capillary capacity and activity. Briefly, it may be assumed that the following changes combine to relieve asthmatic subjects: relief of veinlets in the bronchioles and alveoli; increase of breathing capacity consequent on cardiac shrinkage; elimination of toxins by diuresis and diaphoresis; direct and reflex influences, improved digestion, and subsidence of gastric dilatation; the effects on the general health of better rest at night and of increased ability to take outdoor exercise. Be that as it may, it needs no argument to show that an asthmatic patient is in better case when strong and well-contracting heart-muscles propel the blood-stream through channels which offer a reduced, and perhaps no more than a normal resistance. In this connection I may state that convalescence from acute bronchitis and from pneumonia

---

* Each of these three patients, showing signs of relapse, underwent, after about a year's interval, a course of artificially-prepared baths, with satisfactory results.

may be favourably influenced by recourse to either method, more especially as regards the drying up of moist exudations.

A considerable proportion of other subjects of cardiac dilatation are also affected with frequently recurring distension of the stomach, or with chronic dilatation of that organ. These conditions generally subside *pari passu* with the improvement in the state of the heart without special treatment.

The changes in the general circulation, and more especially in the peripheral vessels and the capillaries, which lead to habitual coldness of the extremities with a deep bluish-red colour of the hands, which gives place to a white hue on pressure, but returns the moment the pressure is relaxed, and not infrequently similarly affect the colour of the cheeks and of the tip of the nose, yield equally well to both baths and exercises where these conditions have not become hopelessly confirmed. Many such cases have been apparently cured. In one an habitual headache, which, with occasional variations of intensity and a few complete intervals, had existed for about seven years, was effectually relieved. The patient was a lady twenty-two years of age. One in whom the treatment produced only partial and temporary relief, and no permanent benefit, was over thirty years of age and also the subject of habitual headache with occasional accesses of acute hemicrania. A man seventy-four years of age, who for four months had been troubled with habitual headache associated with the evacuation of uric acid crystals and accesses of lumbago, was relieved of the headache in four days. The exercises were persevered with for a month, and the apex beat, which had been found an inch outside, receded to a point half an inch within, the nipple line.

Of women who had habitually suffered acutely during the first hours of the menstrual period, a large proportion have reported that they were unconscious of discomfort during the initial stage of the first menstrual period which occurred after either exercises or baths had been commenced. In most of those cases of which I have been able to obtain subsequent information, the relief has been permanent.

The structural changes occurring in the heart and vessels, which are generally designated atheromatous, and which have been regarded as due to the irreparable decay of nature, yield, as already stated, in a manner which is nothing less than surprising to the influences of the baths, but it is only in comparatively early cases that repair can be expected to be carried to the point of completion by one course. Very satisfactory results have been observed in patients who had advanced to the age of seventy-four and upwards. But it has to be borne in mind that subjects of that disease, as of most other affections of the organs of circulation, are the victims of chronic self-poisoning originating in the alimentary canal. It, therefore, follows that dependence must not be placed on physical treatment alone, and that diet and gastro-intestinal antisepsis, as well as measures calculated to ensure effectual elimination, must be regarded as indispensable adjuvants; and it may be stated in this connection that of all internal remedies for cardiac affections generally, aneurysm not excepted, water is perhaps the most powerful and important. It should not be taken in considerable quantities in such relation to meals as to effect injurious dilution of the gastric juice. In power to free the blood, by means of renal excretion, of those toxic ingredients which induce chronic contraction of arterioles and capillaries, and eventually degeneration of structure, and to lower intra-arterial pressure, it stands far in advance of pharmaceutical preparations, which, nevertheless, and more especially iodide of potassium and mercurials, may often be usefully employed as adjuvants.

This brief notice of what may be called the secondary or indirect results of a treatment which is more especially directed to the heart, would be incomplete if I were not to allude to its psychological influence. No one can have observed the subjects of cardiac inefficiency, especially those who are affected by either simple dilatation, or by that condition associated with valvular lesion and failure of compensation, without being struck with the nerve-tension and mental suffering which they endure. Intolerance of sound, irritability, difficulty of mental concentration, lessened power of work, depression amounting, in some cases, to despondency, and night alarm, are of common occurrence. With the rehabilitation of the heart and vessels which these methods of physical treatment are so successful in inducing, all such nerve-suffering vanishes like a dream, and the spirits rise to a plane of hope and energy which is surprising alike to the patient and the physician (Cases E and F, Ch. VIII.).

CHAPTER VII.

THE EXERCISES.

"Movements without design weaken the heart; movements with design, on the contrary, strengthen the heart."—THEO. SCHOTT.

[*For the illustrations contained in this chapter I am indebted to the joint labours of one of my assistants and of Mr. Prendergast Parker, the artist.*—W. B. T.]

IN approaching the subject of the movements which have been shown to exercise therapeutic influences over the heart and blood-vessels, which place the drugs hitherto relied on completely in the shade and relegate them to the position of occasional auxiliaries, it cannot be too clearly stated that we have not to do with "gymnastics" in the sense in which that word is usually employed in the English language. They do, doubtless, in the end, promote the development of the muscles generally, but that is not their primary object. It should be distinctly understood that they are designed to produce regulated movement with *little exertion and no fatigue.*

The person who administers them, who may be called the "operator," should strictly observe and enforce the following rules:—

1. Each movement is to be performed slowly and evenly, that is, at an uniform rate.

2. No movement is to be repeated twice in succession in the same limb or group of muscles.

3. Each single or combined movement is to be followed by an interval of rest.

4. The movements are not to be allowed to accelerate the patient's breathing, and the operator must watch the face for the slightest indications of : (a) dilatation of the alæ nasi ; (b) drawing of the corners of the mouth ; (c) duskiness or pallor of the cheeks and lips ; (d) yawning ; (e) sweating ; and (f) palpitation.

5. The appearance of either of the above signs of distress should be the signal for immediately interrupting the movement in process of execution, and for either supporting the limb which is being moved, or allowing it to subside into a state of rest.

6. The patient must be directed to breathe regularly and uninterruptedly, and should he find any difficulty in doing so, or for any reason show a tendency to hold his breath, he must be instructed to continue counting, in a whisper, during the progress of each movement.

7. No limb or portion of the body of the patient is to be so constricted as to compress the vessels and check the flow of blood.

The following are the movements :—

No. 1.—The arms are to be extended in front of the body on a level with the shoulder joints, the palms of the hands meeting in front of the chest (Fig. 16). The operator places his hands on the outer surface of the patient's wrists in such a manner that the ulnar side of the patient's wrist rests in the fork between his own thumb and forefinger. He places one foot in front of the other so that he may lean forward, without overbalancing himself, while the patient's arms are carried outwards until they are in line with each other, and with the transverse diameter

CHRONIC DISEASES OF THE HEART.   93

of the chest. The operator then places his hands, with a similar disposition of the thumb and forefinger, on the palmar surfaces of the patient's wrist, and

FIG. 16.   FIG. 17.

offers resistance while the arms and hands are being brought back to the position from which they started (Fig. 17).

No. 2.—The arm and hand of one side at a time are extended in the depending position, with the palm of the hand directed forwards, and the operator, standing at the patient's side, places his open hand on the palmar surface of the patient's wrist, the thumb only being on the dorsal surface (Fig. 18). The patient then flexes the forearm, without movement of the upper arm, until the fingers come into contact with the shoulder. The operator then places

FIG. 18.   FIG. 19.

the palmar surface of his own hand on the dorsal surface of the wrist, and maintains it there while the flexed arm is being extended to the position from which the movement commenced (Fig. 19).

No. 3.—The arms are extended vertically in the depending position, with the palms of the hands turned forwards. After they have been raised outwards until the thumbs meet over the head, they are brought back to the original position. The operator faces the patient, and resists the upward movement

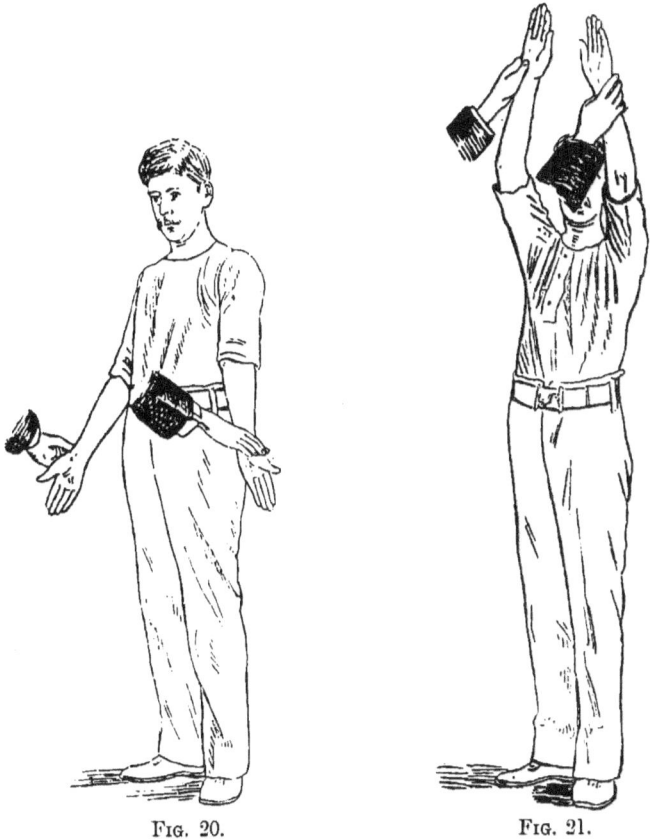

Fig. 20.   Fig. 21.

on the radial side of the wrist (Fig. 20), and the downward movement on the ulnar side (Fig. 21).

No. 4.—The hands, with fingers flexed from the end of the first phalanx in such a manner that the second phalanges of the respective fingers of the two hands are in apposition with their fellows of the opposite side, are pressed together in front of the lower part of the abdomen. The thumbs are extended, and lie within the three sides of a rectangle formed by the flexed forefingers, and touch each other at their tips (Fig. 22). The arms are then raised until the hands are on a level with the vertex of the head. Resistance is offered by placing the hands on the radial surface of the wrists. The movement is then reversed.

Fig. 22.   Fig. 23.

Before the return movement is performed the operator changes the position of his hands so as to receive the wrists in the fork between his thumb and forefinger, the palmar surface of his fingers being applied to the palmar surface of the patient's wrists (Fig. 23).

No. 5.—The extended arms are placed in the depending position, with the palms of the hands resting against the thighs. They are then raised in parallel planes until vertically extended. The movement is then reversed. The operator faces the patient, and in order that he may maintain an uniform and effectual resistance, the relation of his hands to the patient's wrists must pass through the

Fig. 24.

following changes: In the first position the fork between his thumb and forefinger must be applied to the radial part of the wrist (Fig. 24). As the arms rise to an angle of 45° to the body, his fingers glide round the wrist until they are lightly folded round the radial surface of the wrists. Before the reverse movement commences he receives the ulnar aspect of the wrist in the fork between his thumb and forefinger (Fig. 25). While the arms are descending his thumbs move outwards, and at the same time, the fingers glide round the dorsal surface of the wrist in a direction opposite to that which his thumb is taking, in such a manner, and at

Fig. 25.   Fig. 26.

such a rate, that, when the patient's arms are on a level with the shoulders, the ulnar aspect of the wrist rests on a reversed fork formed by the radial aspect of operator's forefingers, and the thumb pushed out to a right angle with the somewhat flexed fingers (Fig. 26). As the hands descend towards the thigh the tips of the operator's fingers gradually glide round to the ulnar aspect of the wrist, so as to resist the downward and backward movement of the arms. This is the operator's *pons asinorum*, but it should be mastered.

No. 6.—The trunk is flexed forward, without the knees being bent, and then brought back to the erect

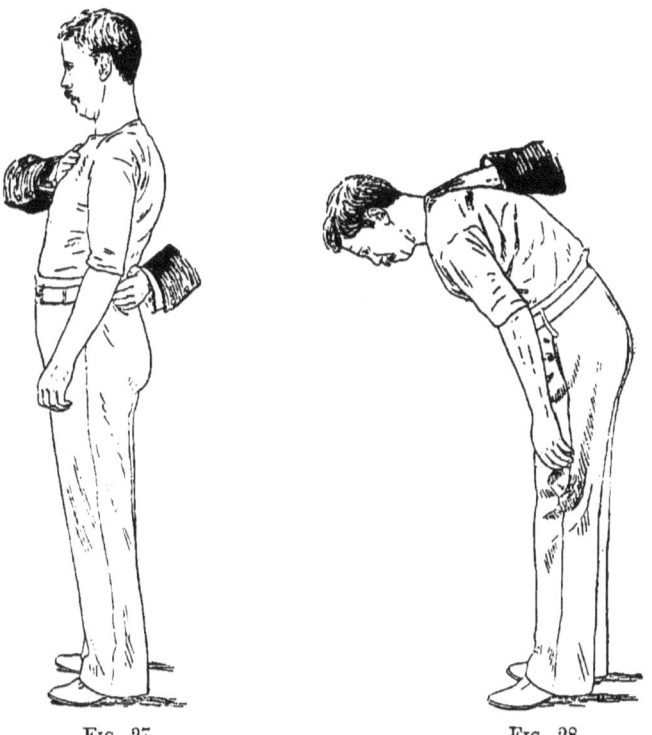

FIG. 27.   FIG. 28.

position. The operator stands at the patient's side with one hand over the upper third of the sternum, and the other supporting the mid-lumbar region (Fig. 27). The reverse movement is resisted by placing one hand over the junction of the cervical and dorsal portions of the spine (Fig. 28).

No. 7.—The trunk is rotated, without movement of the feet, as far as it can be carried to one side, say to the right, then to the left, and lastly brought back to face forwards as at starting. The movements are resisted by one hand being placed in front of, and a little above, the advancing axilla, while the other is

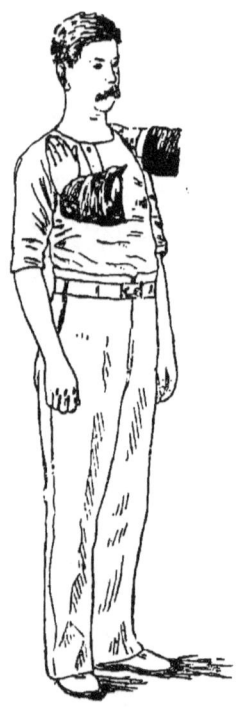

Fig. 29.

placed over the receding shoulder (Fig. 29). The operator must, to a limited extent, move round the patient when the second stage of the rotation is being performed, and will be able to do so most evenly and securely by carrying one foot round behind the other, somewhat as is done in performing the skating "outside edge backwards," before shifting the position of the other.

No. 8.—The trunk is flexed laterally, first to one side, secondly completely over to the other, and thirdly brought back to the erect position. The

Fig. 30.

operator stands in front of the patient. When the movement is to the right, his left hand is pressed against the right side of the chest in the axilla, while the right firmly supports the opposite hip, and *vice versâ* (Fig. 30).

No. 9.—This movement is identical with No. 2, with the exception that while it is being executed the fists are kept firmly clenched.

No. 10.—The arms are flexed in succession as in movement No. 2, with this difference, that the

Fig. 31.

palmar surface is turned outwards and the fist is firmly clenched (Fig. 31).

No. 11.—The arm is extended in the depending position, the palm of the hand lying against the thigh, and then makes a complete revolution from the shoulder joint, forwards and upwards, until it is vertically raised alongside of the ear. Before it descends backwards, the palm of the hand should be turned outwards (Fig. 32). The operator stands at the patient's side with his fingers folded round the

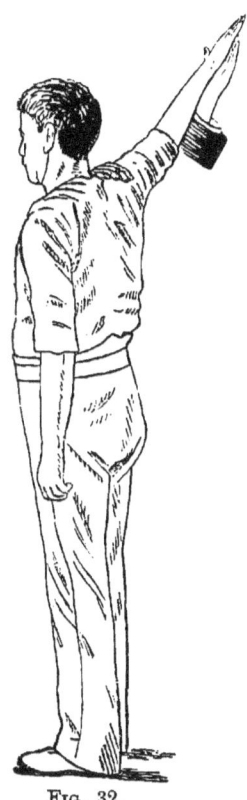

Fig. 32.

radial side of the wrist. His other hand must be ready to receive the wrist when it reaches the vertical position, and to maintain the resistance until the arm has descended to the position from which it started. This movement is performed by one arm at a time.

No. 12.—The arms are extended vertically in the depending position, the palms of the hands resting against the thighs. They are then moved upwards and backwards in parallel planes as far as it is possible

Fig. 33.

to do so without bending the trunk forwards. The upward movement is resisted with the fork of the hand on the ulnar aspect of the wrist, the downward by folding the fingers round the radial surface (Fig. 33).

CHRONIC DISEASES OF THE HEART. 105

No. 13.—The patient stands with one hand resting on a chair or table, while the thigh of the opposite side is flexed on the trunk to the extreme limit, and then extended until the feet are side by side. The

FIG. 34.

leg should hang downwards from the knee-joint. The upward movement is resisted by a hand placed immediately above the knee (Fig. 34). The return

106   THE SCHOTT METHODS OF THE TREATMENT OF

may be resisted by a hand placed below the lower part of the thigh or under the sole of the foot.

No. 14. — The patient, supporting himself with

Fig. 35.

one hand, as in the last movement, bends the whole extended lower extremities in succession, first for-

CHRONIC DISEASES OF THE HEART. 107

wards to the extreme limit of movement, then backwards to the same degree, and finally brings the one foot alongside of the other. The forward movements

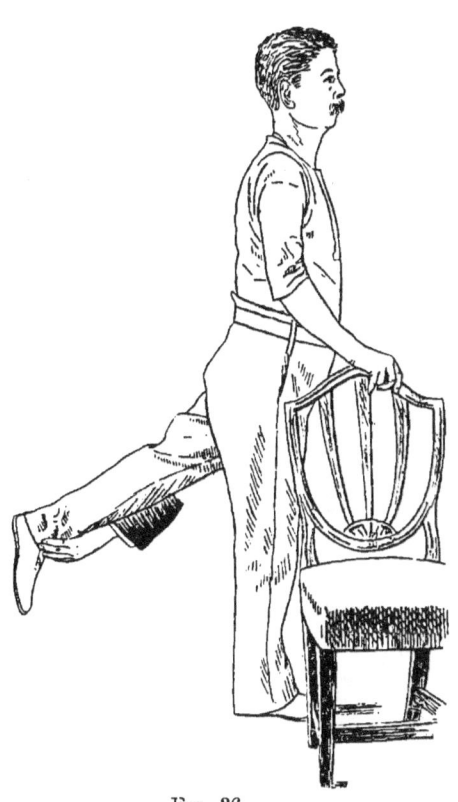

FIG. 36.

are resisted in front of and above the ankle (Figs. 35, 36), the backward movements behind.

No. 15.—The patient, supported in front by a chair or table, stands on either foot in succession,

Fig. 37.

while the leg of the other side is flexed on the thigh. The upward movement is resisted by pressure on the heel (Fig. 37), the return movement above the instep.

No. 16.—The patient, resting one hand on a chair and standing on the foot of the same side, raises the

CHRONIC DISEASES OF THE HEART.   109

extended lower extremities in succession, outwards from the hip joint, and then reverses the movement.

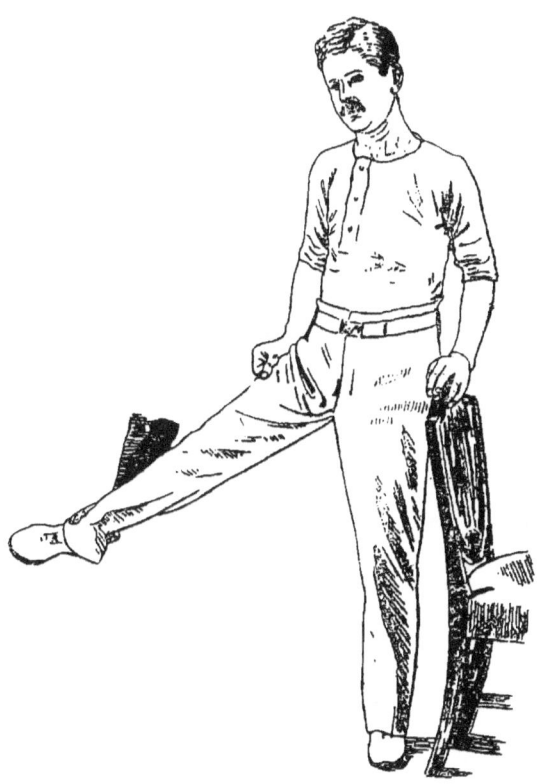

Fig. 38.

The operator resists by means of one hand placed above the ankle (Fig. 38).

No. 17.—The arms, extended horizontally outwards, are rotated from the shoulder-joint to the extreme limits, forwards and backwards. The movements may be resisted by the operator grasping the ulnar edge of the metacarpal portion of the hand

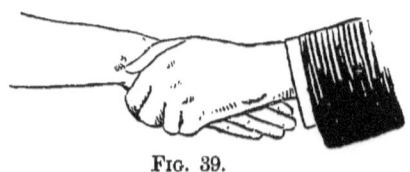

Fig. 39.

(Fig. 39), or by closing his thumb and forefinger in a ring round the wrist.

No. 18.—The hands, in succession, are first extended, then flexed on the forearm to the extreme limits, and lastly brought into line with the arm.

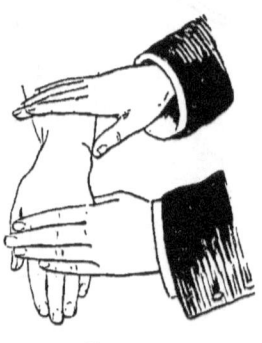

Fig. 40.

The operator's one hand supports the wrist, while the other resists the movements at the metacarpophalangeal junction, first on the dorsal, secondly on the palmar, and thirdly again on the dorsal surface (Fig. 40).

No. 19.—The feet, in succession, are flexed and extended to the extreme limits, and then brought back to their natural position. The movements are resisted in the dorsal and plantar surfaces, at about

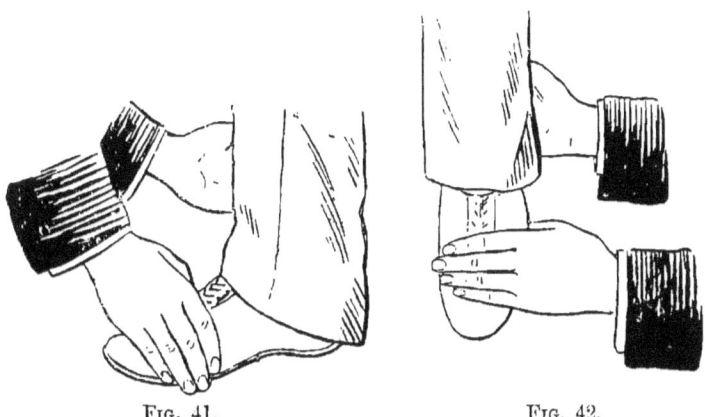

Fig. 41. Fig. 42.

the level of the metatarso-phalangeal joints (Figs. 41, 42).

Such being the mechanical details of the therapeutic movements, it is not surprising that, on making acquaintance with them, the patient asks, with scarcely veiled scepticism, why a wash in the "waters of Jordan" should not be equally effectual. No question could be more apposite, no allusion more appropriate. Many days have not passed before results too manifest to be mistaken offer an unequivocal reply. At the same time, no greater mistake could be made than to assume that the mastery of mechanical details is a sufficient equipment for either the physician or the operator, for in no two cases is their expert and judicious application likely to be precisely similar.

In the first place it should be understood that there

is no magic in the exact sequence which has been adopted in the foregoing description. Many patients are, at the beginning, unable to perform the full series without experiencing what is always to be avoided—namely, fatigue or distress as exhibited by one or more of the symptoms which have been enumerated. Others cannot with advantage submit at once to movements of some special parts, such as the trunk or lower extremities. Some who are confined to bed cannot, in the nature of things, execute a portion of the exercises. In all such respects it devolves on the medical adviser to instruct the operator. The time to be occupied by the several movements, the duration of the interval of rest, and the measure of resistance to be offered, are points on which his judgment should be expressed. For that reason he should always be present when the first exercises are administered, and, in many cases, it will be advisable for him to conduct a few movements himself, and then, having gauged the patient's powers and capacity, to administer them to the assistant in order that his estimate of the required rate of movement and degree of resistance may be placed beyond the possibility of misapprehension. It needs not to be said that the medical attendant should, in all cases, submit the patient to an exhaustive preliminary examination. More especially is this the case where there exists any impediment to the rapid filling of the expanding arteries and capillaries. Should, for example, the pulmonary circuit be obstructed, as it is in cases of emphysema and asthma, and with rigidity or stenosis of the aortic orifice, syncope may be easily induced. In presence of such conditions, the resistance should be limited to feather weight, the movements slowly executed, and the intervals prolonged to allow the heart and vessels time for the adjustment of their mutual relations to the changes which are being rapidly effected in the flow and distribution of the blood. It may even be desirable to enforce the recumbent position lest the pressure in the cerebral vessels be unduly lowered by the imperative requirements of the increasing vascular capacity. If the right side of the heart be overloaded, a down-grade should not, at first, be given to the brachial veins by raising the arms above the level of the shoulders. Briefly, the system under consideration brings such powerful influences to bear on the whole circulation, that, in application, it requires to be adapted to the exact condition of each individual. As with other potent remedies, the "rule of thumb" may easily convert a therapeutic agent into an instrument of mischief. No less care should be taken in the selection, instruction, and supervision of operators. They should be intelligent, light of hand, endowed with powers of observation, and trained to use them. The choice of women is not limited. Trained and even "registered" nurses who possess a knowledge of elementary anatomy and physiology, and whose faculties have been cultivated by hospital service, abound in our country; but suitable men are not easily found. With regard to treatment, although the physical method relegates pharmaceutical remedies to the rank of auxiliaries, their influence is, in some instances, of material value in correcting a special defect of health or in raising the general tone of the system. The patient's daily life often needs regulation, more especially in regard to exercise, fresh air, and the avoidance of undue fatigue, excitement, anxiety, mental distress, and all other depressing conditions. Diet, however, is a matter of scarcely secondary importance. The condition into which most patients have fallen, and the acceleration of tissue change which the bath and exercises alike induce, demand a liberal supply of muscle-forming nourishment, comprising, generally, animal food, though not of necessity butchers' meat, three times daily. If there be a class of subjects with regard to whom the adjustment of the dietary claims exceptional care, it is that numerous one in which a tendency to the excessive deposit of adipose tissue is the accompaniment of anæmia or of some other dyscrasia. With them, the substitution of animal food for a considerable proportion of the fats and carbohydrates in common use, is a measure of great importance. A judiciously devised "thinning," but not "lowering," diet lightens the corporeal burden, gives free play to the muscles, and strengthens the heart to an extent which can hardly be accounted for by the mere removal of superincumbent fat. It should be added that those who have had the widest experience of the Schott methods attach no importance whatever to special limitation of the quantity of fluids ingested, and that graduated mountain climbing, as recommended by Oertel, should only be resorted to towards the end of the treatment or after it has been brought to a satisfactory conclusion. It then forms the rational complement of the treatment. Physical exercise, practised by means of mechanical appliances, forms no part of the system, and introduces principles which are not only foreign to its conception but essentially opposed to it.

It now remains to be said that exercises with "self-imposed resistance" are often found to be of value as an after-treatment, more especially as they are within the competence of everyone who has become acquainted with the movements, and involve no risk of injury by over-exertion. "Selbst-hemmungs-gymnastik" or self-restraining gymnastics, were devised to enable patients to be, if one may so express it, their own operators. The restraint or resistance is effected by that hardening of the muscles of the limbs, or groups of muscles, which execute the movements, of which the condition of the forearm produced by firmly clenching the fist is an example. After a little practice the patient can induce that condition at will, and maintain it throughout the several movements, especially those of the arms and legs.

It would be unbecoming to close these observations without offering a tribute of admiration to the industry and genius which August and Theodor Schott have displayed in devising and elaborating means at once so simple and so effectual for the relief of a large measure of disablement and disease, and, at the same time, acknowledging the generous spirit in which they have, consistently with the most honourable professional ethics, made every effort to bestow the fruit of their labours on the medical profession at large for the benefit of suffering mankind.

CHAPTER VIII.

ILLUSTRATIVE CASES.

C., a lady, aged twenty-seven, had resided for four years in one of the semi-tropical States of America. Had suffered frequent accesses of tertian fever, which throughout the summer of 1893 had continued in unbroken series. Presented intense anæmia, dyspnœa on exertion, and sallow complexion. Suffered continuous headache and chronic intestinal catarrh, to which she had been liable for years. At the termination of the course the anæmia was completely relieved. The headache and dyspnœa were relieved by the end of the first week. The intestinal catarrh was pharmaceutically treated and relieved, but showed a tendency to return on slight provocation. Three years after the completion of the course the improvement was found to have been maintained.

NOTE.—*Areas of cardiac dulness and apex beats indicated by red lines and crosses, respectively, refer to observations made after either baths or exercises.*

DIAGRAM C.

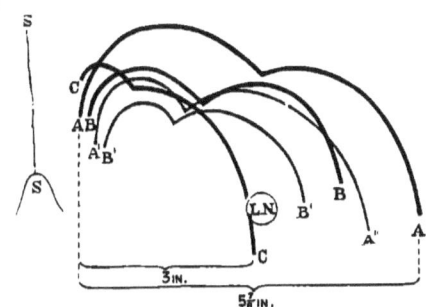

A A and A¹ A¹, areas of cardiac dulness before and after first exercises (1 to 13).
B B and B¹ B¹, areas of cardiac dulness before and after last series of exercises at end of third week.
C C, area of cardiac dulness on termination of four weeks' course.
R.N. and L.N., right and left nipples.
S S, mid-sternal line.

DIAGRAM D.

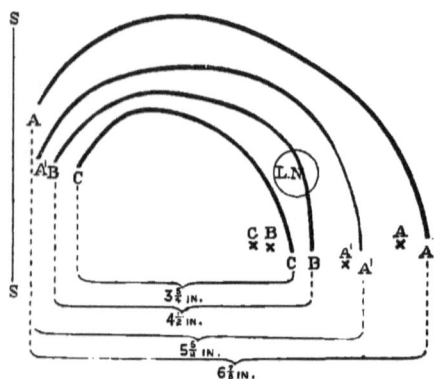

A A, area of cardiac dulness before first exercises.
A¹ A¹, the same after twenty minutes' exercises.
B B, the same after completion of the course, thirty-one days later.
C C, the same eighty days after completion of the course.
× A and ×, positions of apex beat at corresponding stages.
R.N. and L.N., right and left nipples.
S S, mid-sternal line.

D., aged sixty-five, had rarely smoked less than twenty cigars a day during a thirty years' residence in India, and was found to be unduly stout, with pale and drawn face, and light bluish lips. He could not walk a hundred yards without stopping to recover his breath, nor ascend a flight of stairs without resting, supported on the banister, for the same purpose. He was dieted to reduce his weight and correct gastric fermentation, and treated by exercises. Before the completion of the course his aspect and expression had changed, and his face and lips became ruddy; he walked daily to and from his club, a distance of five miles in all, and there played billiards for two or three hours, and could run up stairs without becoming breathless. In weight he lost a pound a week for six weeks. Nine months after the conclusion of the treatment, he was fishing and shooting in Norway, and now smokes, on an average, six small cigars a day. Two years after the commencement of the treatment, the patient was found to be relapsing. A second course of treatment, consisting mainly of baths, yielded equally satisfactory results.

E., aged fifty, had presented symptoms of cardiac failure for at least twenty years, and had been liable for four years to accesses of partial syncope, associated with gastric distension and intense vertigo. The treatment by exercises commenced a few days after recovery from the last such attack. On the conclusion of the course, the patient summarised the change in her condition as follows : "Before the treatment my sensations frequently forced upon me the apprehension of impending death, my digestion was bad, and every exertion of mind and body seemed to be too tiring to be endured. Now I walk for at least an hour twice daily. I can eat and drink anything in reason, and I am a stranger to fatigue and depression of spirits." Four months after the completion of the treatment the improvement was more than maintained. The patient had crossed a mountain pass at an altitude of more than 7,000 feet, in a snowstorm, without the breathing being affected, or experiencing any inconvenience. A year later there had been no recurrence of the symptoms.

DIAGRAM E.

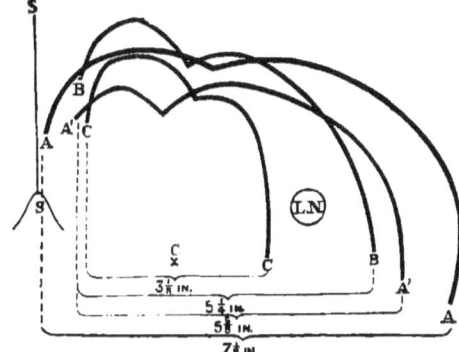

A A and A¹ A¹, areas of dulness before and after exercises. (Pulse reduced from 84 to 76.)
B B. area of dulness after twelfth exercises (the fourth after an interval of fourteen days, necessitated by the menstrual period).
C C. the same after twenty-fifth and last exercises.
C X. situation of apex beat on that occasion (not having been appreciable when previous observations were recorded).
R.N. and L.N., right and left nipples.
S S. mid-sternal line.

To face page 118.

DIAGRAM F.

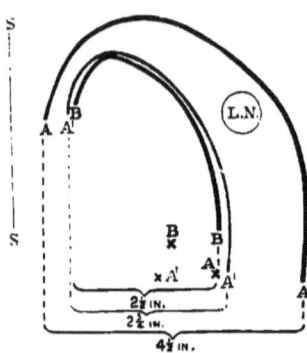

A A and A¹ A¹, areas of cardiac dulness before and after first exercises. (Pulse reduced from 108 to 104.)

B B, area of cardiac dulness twenty hours after completion of the course. (Pulse 84, after going up and down two flights of stairs.)

×A¹, A×, and B×, positions of apex beat at corresponding stages.

R.N. and L.N., right and left nipples.

S S. mid-sternal line.

To face page 119.

F., aged eighteen, 5 ft. 10 in. in height, had a first access of influenza in 1890, and a second in 1892. The latter was followed by loss of ocular accommodation, as well as cardiac weakness and vertigo which became so severe, about three weeks after the termination of the febrile state, that he was obliged to take to his bed, and there lay with a basin at his side, because an incautious movement of the head, or even the automatic fixing of the eyes on a crack in the ceiling, except while wearing convex lenses, brought on an attack of retching. From that time forward he was debarred from participation in all games and sports, as any exertion beyond a leisurely walk brought on palpitation, præcordial pain, and dyspnœa. On the 28th of February, 1894, he commenced a course of exercises which extended to the 24th of March, inclusive, after which he returned to the country. A month later I received the following report:—In active pursuits he is now on a level with other young men of his age. His tutor reports that in power of application, and in memory, he is twice the man he was; but the most remarkable change is in his spirits, for, whereas the word "beastly" used to be freely scattered through his letters, everything in life is now said to be "awfully jolly." After the lapse of a year the patient was in good health and leading an active life. Eighteen months later the patient was reported to be in good health and to be leading an active life. In the winter and spring of 1895—6 he played football and rowed in "College eights," subsequently to suffering an access of influenza. Tachycardia (pulse 110—120) and a measure of dilatation of the right side of the heart ensued. A course of baths restored him to a state of health which has since been maintained.

G.—The following case is published by permission of Dr. James Harper, with whom I saw the patient six weeks after the termination of the acute stage of influenza. She was twenty years of age, and had not gained in strength or power of movement from the time of leaving her bed. She was found to be very anæmic, somewhat wasted, and could only move from one room to another adjoining, at the cost of dyspnœa and præcordial pain. A well-marked systolic bruit was audible at the base. I administered exercises very slowly, with gentle resistance, and long intervals, for the space of fifteen minutes in all. The areas of dulness were traced by Dr. Harper, who also recorded the following observations :—

Dec. 12, 1893.	Before exercises,	P. 96,	murmur distinctly audible.
	After „	P. 84,	„ scarcely „
„ 14, „	Before „		„ fairly marked.
	After 20 m. „		„ scarcely audible.
„ 18, „	Before „	P. 84-120 „	audible.
		(very variable).	
	After 20 m.,,	P. 84,	„ not audible.
„ 28, „	„ „ „	P. 92, no bruit.	
Jan. 1, 1894.	Before „	P. 132,	„
	After 24 m. „	P. 84,	„
„ 4, „	„ „ „	P. 84,	„
„ 9, „	„ ½-hour's „	P. 96,	„
„ 12, „	Before exercises,	P. 96,	„

After the completion of the course the patient travelled by rail from Victoria station to South Kensington, whence she walked to my house. She presented the appearance, and enjoyed the sensations, of perfect health. The pulse was 88, and I could discover no bruit. Up to June, 1898, the patient had remained in excellent health.

DIAGRAM G.

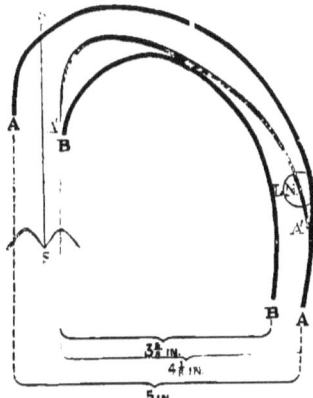

A A and A' A', areas of cardiac dulness before and after fifteen minutes' exercises, with gentle resistance.

B B, the same at the conclusion of a course extending over thirty-one days.

R.N and L.N, right and left nipples.

S S mid-sternal line.

DIAGRAM II.

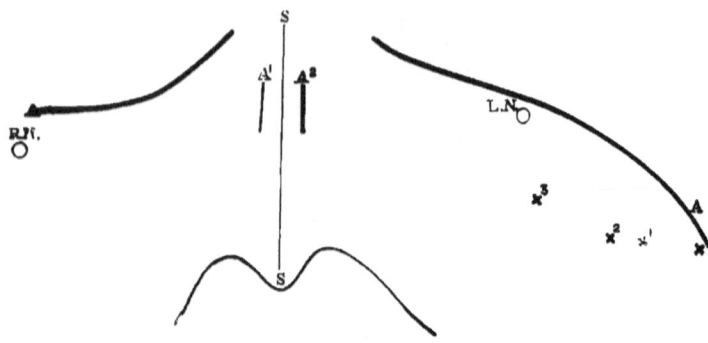

A A, area of dulness before exercises.
A¹, right margin of area of dulness after twenty-five minutes' exercises, with slight resistance.
A², the same, after five more movements, with strong resistance.
×, ×¹, ×², positions of apex beat at corresponding stages.
×³, apex beat as observed by the writer after twenty-five days' treatment by baths and exercises.
R.N. and L.N., right and left nipples.
S S, mid-sternal line.

H., aged fifty-three, was examined in the presence of Dr. Heinemann, of New York, and myself, by Dr. Pawinski, of Warsaw, who kindly sanctions the present use of his observations and tracings. A well-marked systolic bruit at the base, followed by a diastolic souffle in the same situation, were taken to indicate aortic stenosis with insufficiency. The patient suffered from dyspnœa, œdema of the lower extremities, and enlargement of the liver. When I examined him twenty-five days later, the systolic bruit was of diminished intensity, the diastolic souffle was no longer audible, there was no dyspnœa, and the œdema had passed away.

I., aged sixty-seven, had been known for four years to present symptoms of aortic stenosis, but had good compensation and led an active life. He came under observation again on March 8, 1894, some weeks after suffering from symptoms which suggested an attack of influenza. His face was drawn and anxious, and he complained of dyspnœa on exertion, and of great loss of mental and physical energy. On auscultation the basic-systolic bruit was found to have become louder, and to it was superadded a well-marked apex-systolic murmur. The first exercises reduced the pulse from 60 to 50, and increased its force and volume. After seven days (B) he left London much improved, both murmurs being audible but reduced in intensity. Resumed the treatment after an interval of twenty-six days (C), enjoying at the time good general health and complete freedom from dyspnœa. The first exercises of this series reduced the pulse from 72 to 44. In ten more days the treatment was brought to a conclusion by the necessity of leaving London again. By that time the apex bruit had been superseded by a sound which, but for a slight lack of definition, was healthy. Eighty-five days later the area dulness was found to be as indicated by D. The basic bruit was reduced to its old intensity; the apex sound remained as when last observed. The pulse was 52. The general condition left nothing to be desired. The greater part of the members of this patient's family have, in health, a pulse of about 50. Symptoms of relapse were observed in November, 1895, following influenza, and were effectually relieved by a second course consisting mainly of baths. This patient died suddenly of heart failure in 1897.

DIAGRAM 1.

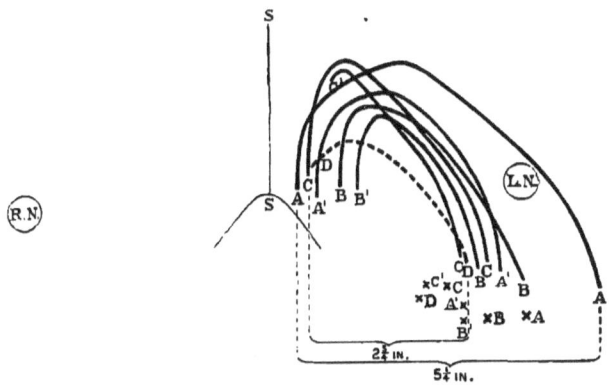

A A and A¹ A¹, areas of dulness before and after first exercises.

B B and B¹ B¹, areas of dulness on seventh day before and after exercises.

C C and C¹ C¹, areas of dulness before and after exercises after an interval of twenty-six days.

D D, area of dulness eighty-five days after conclusion of treatment.

R.N. and L.N., right and left nipples.

S S, mid-sternal line.

× A, × B, etc., positions of apex beat at stages corresponding to letters A, B, C, and D.

× A¹, × B¹, etc., positions of apex beat at stages corresponding to letters A¹, B¹, and C¹.

DIAGRAM K.

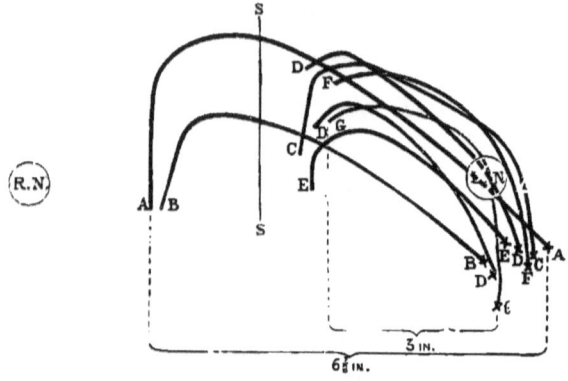

December 7th. A A, before exercises.
B B, after twenty minutes.
,, 9th. C C, before exercises.
,, 12th. D D, before exercises, after being out three and a-half hours and returning in a gale.
D D, after half-an-hour's exercises.
,, 14th. E E, before exercises, and after six and a-half hours' absence from home on professional duty.
,, 25th. F F, before exercises.
January 3rd. G G. three hours after exercises.
R.N. and L.N., right and left nipples.
S S. mid-sternal line.

K., aged fifty-nine, who had for many years been the subject of heart troubles associated with a loud apex-systolic bruit, and exophthalmos, was found on December 3rd, 1893, to be suffering from acute pulmonary apoplexy. The pulse was bigeminous, two beats corresponding to each complete respiratory act. He was at once instructed to practise the arm exercises with self-imposed restraint, and ordered digitalis and strychnia. On the 7th, he was moving about the drawing-room, and commenced a course of exercises resisted by a trained operator. On the 12th, he went to his office and was from home for three and a-half hours. He has since been from time to time under treatment for œdema of the lower extremities, but continued to lead an active professional life until July, 1895, when a serious illness supervened. At the present time the patient is convalescent. The heart is fairly competent and there is no œdema.

NOTE.—This patient died of steadily increasing heart failure, with general anasarca, in 1897.

L., aged sixteen, is reported to have had carditis in the course of scarlet fever, at the age of five, followed by hæmato-albuminuria and œdema. Whooping cough, at the age of six, was followed by chorea of moderate intensity, which lasted for a year. A second attack, lasting four months, occurred at the age of seven. When first seen he was under treatment by Mr. Barwell for spinal curvature. He had a loud systolic-apex bruit, with a well-marked and diffused thrill, and epigastric pulsation perceptible to sight as well as touch. His parents had been advised to remove him from school, and not to allow him to leave the house otherwise than in an invalid chair or pony-chaise. After the exercises on the twenty-first day of treatment he trotted about a hundred and twenty yards and then walked fifty. Before doing so the pulse was 66, and the respirations were 20; afterwards they were respectively 86 and 19. He experienced no fatigue, and showed no signs of distress. At the conclusion of a course extending over twenty-eight days he trotted two hundred yards. Before doing so the pulse was 66, and the respirations were 16; afterwards they were 80 and 18 respectively. He was, at that time, taking walks of one and two hours' duration without fatigue, and, generally, leading an active life, though debarred from running more than a few paces, and from joining in out-door games. The areas of dulness before, and at the conclusion of, the treatment were verified by Mr. Barwell, who also noted a much diminished apex impulse and complete absence of thrill and epigastric pulsation. The bruit had diminished, but was still well marked.

DIAGRAM L.

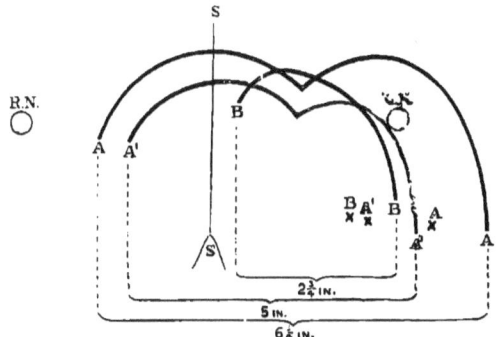

A A and A A¹, areas of cardiac dulness before and after first exercises.
A × and A¹ ×, apex beats before and after.
B B, area of cardiac dulness on completion of course twenty-eight days later.
B ×, apex beat.
R.N. and L.N., right and left nipples.
S S. mid-sternal line.

M.—I am enabled, by the courtesy of Sir Philip C. Smyly, to quote *in extenso* the following case from his article already alluded to :—*

Miss X., æt. seventeen, had been ailing for some time.

Oct. 20th, 1893.—She had an attack of faintness, and swelling of the feet and ankles.

Feb. 24th, 1894.—She came under my care. She was very low and weak; anæmic; hands and feet œdematous and very blue; general anasarca; ascites well marked, and fluid in both pleuræ. The area of the heart dulness was well defined to the right side of the sternum, but could not be outlined below or to the left side on account of the dulness from the pleural effusion. The skin was white and waxy on the forehead, ears, and neck. The cheeks were a dark purple-blue. After a very careful examination the diagnosis arrived at was—Dilated heart with patent foramen ovale (possibly); obstructed arterial circulation, with venous congestion of all the organs. No albumen in the urine. Began the resisted movements for twenty minutes every morning about 11 o'clock, and massage by an experienced masseuse every evening. The colour improved every day, area of dulness diminished, and the pulse became fuller and less frequent. The blue colour improved to a dark red.

March 8th.—In the daily report it was noted :—
" She did some additional exercises with more strength. Her pulse was considerably stronger. Her nose bled slightly. She passed a cheerful day."

11th.—" Marked improvement in the pulse. Her colour keeps good all day; very little blue at any time."

* *Dublin Journal of Medical Science*, September, 1894.

17th.—Remarked the healthier appearance of the forehead, ears, and neck.

(From the 20th of March until the 11th of April the movements were omitted—unavoidably—though the massage was continued.)

21st.—"Not a good night—restless and dreaming. Appetite very poor."

April 5th.—"Her colour was very dark with much blue in the morning; got right in the afternoon. Urine very scanty."

6th.—"Swelling of the abdomen greatly increased. Only eleven ounces of urine in twenty-four hours."

11th.—Pain in the right side. Movements begun again, but very slightly, owing to great distress in breathing. Urine, eleven ounces.

12th.—Pain worse. Much swelling; could not lie down in bed. Urine, thirteen ounces. Daily exercises and steady improvement.

20th.—"Better. Urine, thirty-one ounces in twenty-four hours."

22nd.—"The heart rhythm was normal for the first time."

25th.—"The menses showed for the first time since October, and continued slightly for five or six days."

26th.—The patient was moved from one house to another. On being lifted into the carriage she became breathless and very blue. Towards evening she breathed better, but could not lie down. She got little or no sleep, and had to be supported sitting up all night.

28th.—The whole of the right pleura was full. Distress of the breathing rapidly increasing. Assisted by Dr. Cruise, I tapped the chest and drew off sixty-two ounces of fluid, clear and yellow. She coughed up a quantity of thin mucus during the afternoon. Temperature 100°. At 10 o'clock, p.m., the temperature was 99°. She could lie down in bed with only two pillows.

May 4th.—Consultation with Dr. Cruise. She was very much improved. Movements were resumed.

11th.—" Was very cheerful all day. Colour a little high, but no blue."

21st.—Left Dublin, 7 p.m., for Holyhead. Next day to London, and on Wednesday had a consultation with Dr. Bezly Thorne.

26th.—Arrived at Nauheim better than when she left Dublin.

28th.—Had her first bath.

June 1st.—Pulse before the bath, 116; after, 110.

8th.—Dr. Schott showed the patient's mother "that the water had gone down a hand's breadth over her stomach."

20th.—Began the gymnastics. From standing during the examination, and the marking out the area of dulness, her pulse was 114; after the exercises it fell to 88, and then rose to 96, and then to 104.

21st.—I saw the patient at Nauheim with Dr. Th. Schott. I could not find any sign of fluid either in the chest or in the peritoneum. No œdema; and the heart's action normal.

July 6th.—Dr. Schott reports the heart very well.

29th.—Dr. Bezly Thorne saw the patient in London, and reports—" Wonderfully improved. Cannot detect any wrong sound in her heart."

30th.—Dr. Cruise and I very carefully examined the patient together, and found the heart's action normal, and no swelling anywhere. She returned home to all appearance quite well.

At an early part of this case it was mentioned that it appeared possible, from the extreme cyanosis, that some patency of the foramen existed. It is, however, quite possible that this may have been remedied by the contraction of the dilated heart, and consequent valvular closure of the foramen.

"93, Merrion Square,
"August 9th, 1894.

"Dear Sir Philip,

"Having read your paper on the treatment of enlarged heart by movements of the system of the brothers Schott, I beg to add a short note, which, if you wish, you can publish.

"As you know, I saw your patient in the most critical portion of her illness, and learned, for the first time, what can be done by these movements, and in confirmation of what you succeeded in accomplishing in a young girl, I now beg to report my own experience of the treatment in a very aged patient.

"Within the last month I met Dr. O'Donoghoe, of Baldoyle, in relation to the case of a very aged gentleman, who was suffering extreme distress of breathing and loss of sleep from a weak dilated heart.

"In addition to the administration of iron and digitalis, and stimulation by a small blister, I used the resisted movements of the arms, and I showed them to Dr. O'D., who fully appreciated their object and value. He and some of the patient's family have still continued them, and the patient is totally changed for the better—sleeps well, has recovered his appetite, and physically shows increased impulse and diminished area of cardiac dulness.

"I remain, my dear Sir Philip,
"Yours most faithfully,
"F. R. Cruise."

On the 19th of September Sir Philip reported:—
"X. is wonderful—riding, driving, and boating."
He draws attention to the following points, which this case brings into prominence :

1. The importance of the movements without the baths, followed by such an improvement that the patient was able to undertake the journey to Nauheim.

2. The value of the Nauheim baths in removing the ascites and pleural effusion.

3. The interesting observation that the symptoms returned when the movements were discontinued for some weeks, though massage was continued regularly. In November, 1895, I carefully examined this patient. The cardiac sounds and area of dulness were normal. She was in excellent health, had gained a stone in weight, and was leading an active life. Still later information (May, 1898) showed the patient to be in excellent health.

W. B. T.

INDEX.

Anæmia, 84
Analyses of waters, 12, 13
Aneurysm, 47
Angina cum dolore, 52, 54
Angina sine dolore, 50
Anginous distress in the bath, 39
Animal food, 114
Antisepsis, gastro-intestinal, 89
Aortic disease with mitral, 60
Aortic stenosis, 112
Aortic valves, lesion of, 59
Artificially prepared baths, 21, 22
Asthma, 85, 86, 87
Atheroma, 84, 89
Athlete's heart, 69

Bath waters, 12, 13, 21, 22
Body, weight diminution of, 56
Bronchitis, 87

Carbonic acid, 20, 21
Cold extremities, 88
Contra-indications which are only apparent, 40
Cutaneous excitation, 41, 43

Degeneration myocardial, 71
Diagnostic value of exercises, 47
Diaphragm, level of, 36
Diet, 89
Dilatation, diagnosis of early stages, 47
Disturbing causes interfering with treatment, 72—78
Diuresis, 19, 32
Dysmenorrhœa, 89

Effervescence, production of, 20, 21
Exercises, description of, 91

Exercises, diagnostic value of, 47
  in obstruction of pulmonary circuit, 112
  influence of, 32
  resistance in administering, 112
  with self-imposed resistance, 114

Fluids, limitation of, 114

Gastric dilatation, 88
Graves' disease, 81

Hands, pale or blue, 33, 88
Headache, 88
Heart, contraction of the, 23, 24, 28
Hypertrophy, differential diagnosis of, 47

Inflation of lungs, 35
Influenza, 48
  cardiac sequelæ of, 48, 84, 119, 120
Irresponsible treatment, 45
Isolation, benefit of, 72, 80, 81

Kidneys, effect of baths on, 19
  effects of exercises on, 32

Mechanical appliances, 114
Mitral disease, 56
Mitral, aortic disease with, 60
Mountain climbing, 114
Murmurs, changes in, 32, 120
Myocardial degeneration, 71

Nervous system, influence upon reaction, 42

INDEX.

Œdema, 85, 125, 127

Pneumonia, 87
Prognostic value of exercises, 47
Psychological influence of Schott methods, 90
Pulse, influence of baths on, 17, 18, 23, 24, 25
   influence of exercises on, 27, 28, 34
   pressure, diminution of, 50, 56, 72, 73

Reaction, 40, 44
Recumbent position, 43, 113
Repetition of treatment, 62
Resistance, measure of, 42, 112
Respiration, effects of baths on, 18
Respiratory bruit, 70
Restorative influence of exercises on syncope, 52
Rheumatism, sequelæ of, 83
Rheumatoid arthritis, 84
Right heart, affection of, 66

Smoker's heart, 117
Sprudel bath, 14, 15
   spring, 12
Syncope, 43
   precautions against, 112
   restorative influences of exercises on, 52
Systolic bruit, right heart, 66

Temperatures of natural baths, 12, 13, 40
Trophic influence of baths, 19

Uric acid pains, 66
Urine, increased flow of, 19, 32

Vascular capacity, increase of, 19, 33
   system, effects on, 18, 27, 28, 33, 34, 81, 82, 87
   repair, 81

Warmth in delayed reaction, 44
Water, drinking of, 89
Widerstandsgymnastik, 29

No. 3.

London, 7, Great Marlborough Street,
June, 1900.

A SELECTION

FROM

J. & A. CHURCHILL'S CATALOGUE,

COMPRISING

MOST OF THE RECENT WORKS PUBLISHED BY THEM.

N.B.—J. & A. Churchill's larger Catalogue, which contains over 600 works, with a Complete Index to their Subjects, will be sent on application.

Human Anatomy:
A Treatise by various Authors. Edited by HENRY MORRIS, M.A., M.B. Lond., F.R.C.S., Senior Surgeon to, and Lecturer on Surgery at, the Middlesex Hospital. Second Edition. Roy. 8vo, with 790 Illustrations, nearly all original, and many of them in several colours, 36s.

Heath's Practical Anatomy:
A Manual of Dissections. Eighth Edition. Edited by WILLIAM ANDERSON, F.R.C.S., Surgeon and Lecturer on Anatomy at St. Thomas's Hospital, Examiner in Anatomy for R.C.P. and S. Crown 8vo, with 329 Engravings, 15s.

Wilson's Anatomist's Vade-Mecum. Eleventh Edition. By HENRY E. CLARK, M.R.C.S. Eng., F.F.P.S. Glasg., Examiner in Anatomy, F.P.S., and Professor of Surgery in St. Mungo's College, Glasgow. Crown 8vo, with 492 Engravings and 26 Coloured Plates, 18s.

An Atlas of Human Anatomy.
By RICKMAN J. GODLEE, M.S., F.R.C.S., Surgeon and late Demonstrator of Anatomy, University College Hospital. With 48 Imp. 4to Plates (112 figures), and a volume of Explanatory Text. 8vo, £4 14s. 6d.

Human Osteology.
By LUTHER HOLDEN, Consulting Surgeon to St. Bartholomew's Hospital. Eighth Edition, edited by CHARLES STEWART, F.R.S., Conservator of the Museum R.C.S., and ROBERT W. REID, M.D., F.R.C.S., Regius Professor of Anatomy in the University of Aberdeen. 8vo, with 59 Lithographic Plates and 74 Engravings. 16s.

Also,

Landmarks, Medical and Surgical. Fourth Edition. 8vo, 3s. 6d.

The Student's Guide to Surgical Anatomy. By EDWARD BELLAMY, F.R.C.S. Third Edition. Fcap. 8vo, with 81 Engravings. 7s. 6d.

Anatomy of the Joints of Man.
By HENRY MORRIS, Senior Surgeon to the Middlesex Hospital. With 44 Plates (several coloured). 8vo, 16s.

A Manual of General Pathology, for Students and Practitioners. By W. S. LAZARUS-BARLOW, B.A., M.D., Pathologist and Lecturer on Pathology, Westminster Hospital. 8vo, 21s.

Pathological Anatomy of Diseases. Arranged according to the nomenclature of the R.C.P. Lond. By NORMAN MOORE, M.D., F.R.C.P., Assistant Physician to St. Bartholomew's Hospital. Fcap. 8vo, with 111 Engravings, 8s. 6d.

A Manual of Clinical and Practical Pathology. By W. E. WYNTER, M.D., M.R.C.P., Assistant Physician to the Middlesex Hospital, and F. J. WETHERED, M.D., F.R.C.P., Assistant Physician to the Consumption Hospital, Brompton. With 4 Coloured Plates and 67 Engravings. 8vo, 12s. 6d.

General Pathology:
An Introduction to. By JOHN BLAND SUTTON, F.R.C.S., Assistant Surgeon to the Middlesex Hospital. 8vo, with 149 Engravings, 14s.

The Pathologist's Handbook:
A Manual of the Post-mortem Room. By T. N. KELYNACK, M.D., late Demonstrator in Morbid Anatomy, Owens College, Manchester. With 126 Illustrations, fcap. 8vo, pegamoid, 4s. 6d.

A Contribution to the History of the Respiration of Man; being the Croonian Lectures delivered before the Royal College of Physicians in 1895, with supplementary considerations of the methods of inquiry and analytical results. By WILLIAM MARCET, M.D., F.R.C.P., F.R.S. With Diagrams, imp. 8vo, 5s. 6d.

LONDON: 7, GREAT MARLBOROUGH STREET.

J. & A. CHURCHILL'S RECENT WORKS.

Atlas of the Central Nervous System. From the larger work of Hirschfeld and Léveillé. Edited by HOWARD H. TOOTH, M.D. F.R.C.P. With 37 Plates carefully coloured by Hand. Large Imp. 8vo, 40s.

The Human Brain: Histological and Coarse Methods of Research for Students and Asylum Medical Officers. By W. BEVAN LEWIS, Medical Superintendent, West Riding Asylum. 8vo, with Engravings and Photographs, 8s.

The Physiology and the Pathology of the Cerebral Circulation: an Experimental Research. By LEONARD HILL, M.B., Hunterian Professor, R.C.S. With 41 Illustrations, Royal 8vo, 12s.

Elements of Human Physiology. By ERNEST H. STARLING, M.D., F.R.C.P., F.R.S., Jodrell Professor of Physiology in University College, London. Third Edition. Crown 8vo, with 140 Engravings, 7s. 6d.

Manual of Physiology: For the use of Junior Students of Medicine. By GERALD F. YEO, M.D., F.R.S. Third Edition. Crown 8vo, with 254 Engravings (many figures), and Plate of Spectra, 14s.

A Class Book of Practical Physiology, including Histology, Chemical and Experimental Physiology. By DE BURGH BIRCH, M.D., F.R.S.E., Professor of Physiology in the Yorkshire College of the Victoria University. With 62 Illustrations, crown 8vo, 6s. 6d.

Practical Lessons in Elementary Biology, for Junior Students. By PEYTON T. B. BEALE, F.R.C.S., Lecturer on Elementary Biology and Demonstrator in Physiology in King's College, London. Crown 8vo, 3s. 6d.

Medical Jurisprudence: Its Principles and Practice. By ALFRED S. TAYLOR, M.D., F.R.C.P., F.R.S. Fourth Edition, by THOMAS STEVENSON, M.D., F.R.C.P., Lecturer on Medical Jurisprudence at Guy's Hospital. 2 vols. 8vo, with 189 Engravings, 31s. 6d.

Hygiene and Public Health. A Treatise by various Authors. Edited by THOMAS STEVENSON, M.D., F.R.C.P., Lecturer on Chemistry and Medical Jurisprudence at Guy's Hospital; Official Analyst to the Home Office; and SHIRLEY F. MURPHY, Medical Officer of Health of the County of London. In 3 vols., royal 8vo, fully Illustrated. Vol. I., 28s.; Vol. II., 32s.; Vol. III., 20s.

The Theory and Practice of Hygiene. By J. LANE NOTTER, M.D., Examiner in Hygiene and Public Health in the University of Cambridge and in the Victoria University, Professor of Hygiene in the Army Medical School; and R. H. FIRTH, F.R.C.S., Assistant Professor of Hygiene in the Army Medical School. With numerous Illustrations, Royal 8vo, 24s.

A Manual of Practical Hygiene. By the late E. A. PARKES, M.D., F.R.S. Eighth Edition, by J. LANE NOTTER, A.M., M.D. 8vo, with 10 Plates and 103 Engravings, 18s.

A Handbook of Hygiene and Sanitary Science. By GEO. WILSON, M.A., M.D., LL.D., F.R.S.E., D.P.H. Camb., Medical Officer of Health for Mid-Warwickshire. Eighth Edition. Post 8vo, with Engravings, 12s. 6d.

A Simple Method of Water Analysis, especially designed for the use of Medical Officers of Health. By JOHN C. THRESH, M.D.Vic., D.Sc. Lond., D.P.H.Camb., Medical Officer of Health for the County of Essex. Second Edition. Fcap. 8vo, 2s. 6d.

Elements of Health: an Introduction to the Study of Hygiene. By LOUIS C. PARKES, M D., D.P.H. Lond., Lecturer on Public Health at St. George's Hospital. Post 8vo, with 27 Engravings, 3s. 6d.

Diet and Food considered in relation to Strength and Power of Endurance, Training and Athletics. By ALEXANDER HAIG, M.D., F.R.C.P. Second Edition. Crown 8vo, 2s.

Effects of Borax and Boracic Acid on the Human System. By Dr. OSCAR LIEBREICH, Professor in the University of Berlin. With Plates, post 4to, 2s.

The Prevention of Epidemics and the Construction and Management of Isolation Hospitals. By ROGER MCNEILL, M.D. Edin., D.P.H. Camb., Medical Officer of Health for the County of Argyll. 8vo. With several Hospital Plans, 10s. 6d.

Hospitals and Asylums of the World; their Origin, History, Construction, Administration, Management, and Legislation. By Sir HENRY BURDETT, K.C.B. In 4 vols., with Portfolio. Complete, 168s. Vols. I. and II.—Asylums; 90s. Vols. III. and IV.—Hospitals, &c., with Portfolio of Plans, 120s.

LONDON: 7, GREAT MARLBOROUGH STREET.

A Manual of Bacteriology, Clinical and Applied. With an Appendix on Bacterial Remedies, &c. By RICHARD T. HEWLETT, M.D., M.R.C.P., D.P.H. Lond., Assistant in the Bacteriological Department, Jenner Institute of Preventive Medicine. With 75 Illustrations, post 8vo, 10s. 6d.

Mental Diseases: Clinical Lectures. By T. S. CLOUSTON, M.D., F.R.C.P. Edin., Lecturer on Mental Diseases in the University of Edinburgh. Fifth Edition. Crown 8vo, with 19 Plates, 14s.

A Text-Book on Mental Diseases, for Students and Practitioners of Medicine. By THEODORE H. KELLOGG, M.D., late Medical Superintendent of Willard State Hospital, U.S.A. With Illustrations, 8vo., 25s.

Mental Physiology, especially in its Relation to Mental Disorders. By THEO. B. HYSLOP, M.D., Resident Physician and Medical Superintendent, Bethlem Royal Hospital; Lecturer on Mental Diseases, St. Mary's Hospital Medical School. 8vo, 18s.

The Insane and the Law: a Plain Guide for Medical Men, Solicitors, and Others as to the Detention and Treatment, Maintenance, Responsibility, and Capacity either to give evidence or make a will of Persons Mentally Afflicted. With Hints to Medical Witnesses and to Cross-Examining Counsel. By G. PITT-LEWIS, Q.C., R. PERCY SMITH, M.D., F.R.C.P., late Resident Physician, Bethlem Hospital, and J. A. HAWKE, B.A., Barrister-at-Law. 8vo, 14s.

A Dictionary of Psychological Medicine, giving the Definition, Etymology, and Synonyms of the Terms used in Medical Psychology; with the Symptoms, Treatment, and Pathology of Insanity; and THE LAW OF LUNACY IN GREAT BRITAIN AND IRELAND. Edited by D. HACK TUKE, M.D., LL.D., assisted by nearly 130 Contributors. 2 vols., 1,500 pages, royal 8vo, 42s.

The Mental Affections of Children, Idiocy, Imbecility, and Insanity. By WM. W. IRELAND, M.D. Edin., formerly Medical Superintendent of the Scottish Institution for the Education of Imbecile Children. Second Edition. With 21 Plates, 8vo, 14s.

Mental Affections of Childhood and Youth (Lettsomian Lectures for 1887, &c.). By J. LANGDON-DOWN, M.D., F.R.C.P., Consulting Physician to the London Hospital. 8vo, 6s.

The Journal of Mental Science. Published Quarterly, by Authority of the Medico-Psychological Association. 8vo, 5s.

Manual of Midwifery: Including all that is likely to be required by Students and Practitioners. By A. L. GALABIN, M.D., F.R.C.P., Obstetric Physician to Guy's Hospital. Fourth Edition. Crown 8vo, with 271 Engravings, 15s.

The Practice of Midwifery: A Guide for Practitioners and Students. By D. LLOYD ROBERTS, M.D., F.R.C.P., Consulting Obstetric Physician to the Manchester Royal Infirmary, Physician to St. Mary's Hospital. Fourth Edition. Crown 8vo, with 2 Coloured Plates and 226 Woodcuts, 10s. 6d.

A Short Practice of Midwifery, embodying the Treatment adopted in the Rotunda Hospital, Dublin. By HENRY JELLETT, M.D., B.A.O. Dub., late Assistant Master, Rotunda Hospital. Second Edition. With 57 Illustrations, crown 8vo, 6s.

Obstetric Aphorisms: For the Use of Students commencing Midwifery Practice. By JOSEPH G. SWAYNE, M.D. Tenth Edition. Fcap. 8vo, with 20 Engravings, 3s. 6d.

Economics, Anæsthetics, and Antiseptics in the Practice of Midwifery. By HAYDN BROWN, L.R.C.P., L.R.C.S. Fcap. 8vo, 2s. 6d.

Lectures on Obstetric Operations: A Guide to the Management of Difficult Labour. By ROBERT BARNES, M.D., F.R.C.P. Fourth Edition. 8vo, with 121 Engravings, 12s. 6d.

By the same Author.

A Clinical History of Medical and Surgical Diseases of Women. Second Edition. 8vo, with 181 Engravings, 28s.

Manual of the Diseases peculiar to Women. By JAMES OLIVER, M.D., F.R.S. Edin., M.R.C.P. Lond., Physician to the Hospital for Diseases of Women, London. Fcap. 8vo, 3s. 6d.

By the same Author.

Abdominal Tumours and Abdominal Dropsy in Women. Crown 8vo, 7s. 6d.

Gynæcological Operations: (Handbook of). By ALBAN H. G. DORAN, F.R.C.S., Surgeon to the Samaritan Hospital. 8vo, with 167 Engravings, 15s.

The Student's Guide to the Diseases of Women. By ALFRED L. GALABIN, M.A., M.D., F.R.C.P., Obstetric Physician to Guy's Hospital. Fifth Edition. Fcap. 8vo, with 142 Engravings, 8s. 6d.

A Practical Treatise on the Diseases of Women. By T. GAILLARD THOMAS, M.D. Sixth Edition, by PAUL F. MUNDÉ, M.D. Roy. 8vo, with 347 Engravings, 25s.

LONDON : 7, GREAT MARLBOROUGH STREET.

A First Series of Fifty-four Consecutive Ovariotomies, with Fifty-three Recoveries. By A. C. BUTLER-SMYTHE, F.R.C.P. Edin., Surgeon to the Samaritan Free Hospital, Senior Surgeon to the Grosvenor Hospital for Women and Children. 8vo, 6s. 6d.

Sterility. By ROBERT BELL, M.D., F.F.P. & S., Senior Physician to the Glasgow Hospital for Diseases peculiar to Women. 8vo, 5s.

Notes on Gynæcological Nursing. By JOHN BENJAMIN HELLIER, M.D., M.R.C.S., Surgeon to the Hospital for Women, &c., Leeds. Cr. 8vo, 1s. 6d.

A Manual for Hospital Nurses and others engaged in Attending on the Sick, with a Glossary. By EDWARD J. DOMVILLE, Surgeon to the Devon and Exeter Hospital. Eighth Edition. Crown 8vo, 2s. 6d.

A Short Manual for Monthly Nurses. By CHARLES J. CULLINGWORTH, M.D., F.R.C.P., Obstetric Physician to St. Thomas's Hospital. Revised by M. A. ATKINSON. Fourth Edition. Fcap. 8vo, 1s. 6d.

Lectures on Medicine to Nurses. By HERBERT E. CUFF, M.D., F.R.C.S., Superintendent, North Eastern Fever Hospital, London. Second Edition. With 29 Illustrations, 3s. 6d.

Antiseptic Principles for Nurses. By C. E. RICHMOND, F.R.C.S. 8vo, 1s.

The Diseases of Children. By JAS. F. GOODHART, M.D., F.R.C.P., Consulting Physician to the Evelina Hospital for Children. Sixth Edition, with the assistance of G. F. STILL, M.D., Medical Registrar to the Gt. Ormond St. Hospital for Children. Crown 8vo, 10s. 6d.

A Practical Treatise on Disease in Children. By EUSTACE SMITH, M.D., F.R.C.P., Physician to the King of the Belgians, and to the East London Hospital for Children, &c. Second Edition. 8vo, 22s.

By the same Author.

Clinical Studies of Disease in Children. Second Edition. Post 8vo, 7s. 6d.

Also.

The Wasting Diseases of Infants and Children. Sixth (Cheap) Edition. Post 8vo, 6s.

On the Natural and Artificial Methods of Feeding Infants and Young Children. By EDMUND CAUTLEY, M.D., Physician to the Belgrave Hospital for Children. Crown 8vo, 7s. 6d.

Materia Medica, Pharmacy, Pharmacology, and Therapeutics. By W. HALE WHITE, M.D., F.R.C.P., Physician to, and Lecturer on Pharmacology and Therapeutics at, Guy's Hospital. Fourth Edition, based upon the B.P. of 1898. Fcap. 8vo, 7s. 6d.

An Introduction to the Study of Materia Medica, designed for Students of Pharmacy and Medicine. By HENRY G. GREENISH, F.I.C., F L.S., Professor of Materia Medica to the Pharmaceutical Society. With 213 illustrations, 8vo, 15s.

Materia Medica And Therapeutics. By CHARLES D. F. PHILLIPS, M.D., F.R.S. Edin. Vegetable Kingdom — Organic Compounds — Animal Kingdom. 8vo, 25s. Inorganic Substances. Second Edition. 8vo, 21s.

Recent Materia Medica, and Drugs Occasionally Prescribed; Notes on their Origin and Therapeutics. By F. HARWOOD LESCHER, F.C.S., Pereira Medallist. Fifth Edition. 8vo, 4s.

Practical Pharmacy: An Account of the Methods of Manufacturing and Dispensing Pharmaceutical Preparations; with a chapter on the Analysis of Urine. By E. W. LUCAS, F.C.S., Examiner at the Pharmaceutical Society. With 283 Illustrations. Royal 8vo, 12s. 6d.

Galenic Pharmacy: A Practical Handbook to the Processes of the British Pharmacopœia. By R. A. CRIPPS, M.P.S. 8vo, with 76 Engravings, 8s. 6d.

Practical Pharmacy. By BARNARD S. PROCTOR, formerly Lecturer on Pharmacy at the College of Medicine, Newcastle-on-Tyne. Third Edition. 8vo, with 44 Wood Engravings and 32 Lithograph Fac-Simile Prescriptions, 14s.

A Companion to the British Pharmacopœia. By PETER SQUIRE. Revised by PETER WYATT SQUIRE. Seventeenth Edition. 8vo, 12s. 6d.

By the same Authors.

The Pharmacopœias of thirty of the London Hospitals, arranged in Groups for Comparison. Seventh Edition. Fcap. 8vo. 6s.

J. & A. CHURCHILL'S RECENT WORKS.

The Galenical Preparations of the British Pharmacopœia: A Handbook for Medical and Pharmaceutical Students. By CHARLES O. HAWTHORNE, M.D., C.M., late Lecturer on Materia Medica, Queen Margaret College, Glasgow. 8vo, 4s. 6d.

The Pharmaceutical Formulary; a Synopsis of the British and Foreign Pharmacopœias. By HENRY BEASLEY. Twelfth Edition (B.P. 1898), by J. OLDHAM BRAITHWAITE. 18mo, 6s. 6d.

By the same Author.

The Druggist's General Receipt-Book. Tenth Edition. 18mo, 6s. 6d.

Also,

The Book of Prescriptions: Containing upwards of 3,000 Prescriptions from the Practice of the most eminent Physicians and Surgeons, English and Foreign. Seventh Edition. 18mo, 6s. 6d.

Selecta è Prescriptis: Containing Terms, Phrases, Contractions and Abbreviations used in Prescriptions, with Explanatory Notes, &c. Also, a Series of Abbreviated Prescriptions with Translations and Key. By J. PEREIRA, M.D., F.R.S. Eighteenth Edition, by JOSEPH INCE, F.C.S., F.L.S. 24mo, 5s.

Year-Book of Pharmacy: Containing the Transactions of the British Pharmaceutical Conference. Annually. 8vo, 10s.

Southall's Organic Materia Medica: a Handbook treating of some of the more important of the Animal and Vegetable Drugs made use of in Medicine, including the whole of those contained in the B.P. Fifth and Enlarged Edition. By JOHN BARCLAY, B.Sc.Lond., some time Lecturer on Materia Medica and Pharmacy in Mason College, Birmingham. 8vo, 6s.

Manual of Botany. By J. REYNOLDS GREEN, Sc.D., M.A., F.R.S., Professor of Botany to the Pharmaceutical Society. Two Vols. Cr. 8vo. Vol. I.—Morphology and Anatomy. Second Edition. With 778 Engravings, 7s. 6d.
„ II.—Classification and Physiology. With 417 Engravings, 10s.

The Student's Guide to Systematic Botany, including the Classification of Plants and Descriptive Botany. By ROBERT BENTLEY. Fcap. 8vo, with 350 Engravings, 3s. 6d.

Medicinal Plants: Being descriptions, with original figures, of the Principal Plants employed in Medicine, and an account of their Properties and Uses. By Prof. BENTLEY and Dr. H. TRIMEN, F.R.S. In 4 vols., large 8vo, with 306 Coloured Plates, bound in Half Morocco, Gilt Edges, £11 11s.

Therapeutic Electricity and Practical Muscle Testing. By W. S. HEDLEY, M.D., in charge of the Electro-therapeutic Department of the London Hospital. With 110 Illustrations. Royal 8vo, 8s. 6d.

Practical Therapeutics: A Manual. By EDWARD J. WARING, C.I.E., M.D., F.R.C.P., and DUDLEY W. BUXTON, M.D., B.S. Lond. Fourth Edition. Crown 8vo, 14s.

By the same Author.

Bazaar Medicines of India, And Common Medical Plants: With Full Index of Diseases, indicating their Treatment by these and other Agents procurable throughout India, &c. Fifth Edition. Fcap. 8vo, 5s.

Climate and Fevers of India, with a series of Cases (Croonian Lectures, 1882). By Sir JOSEPH FAYRER, K.C.S.I., M.D. 8vo, with 17 Temperature Charts, 12s.

By the same Author.

The Natural History and Epidemiology of Cholera: Being the Annual Oration of the Medical Society of London, 1888. 8vo, 3s. 6d.

Psilosis or "Sprue," its Nature and Treatment; with Observations on various Forms of Diarrhœa acquired in the Tropics. By GEORGE THIN, M.D. Second and enlarged Edition, with Illustrations. 8vo, 10s.

A Manual of Family Medicine and Hygiene for India. Published under the Authority of the Government of India. By Sir WILLIAM J. MOORE, K.C.I.E., M.D., late Surgeon-General with the Government of Bombay. Sixth Edition. Post 8vo, with 71 Engravings, 12s.

By the same Author.

A Manual of the Diseases of India: With a Compendium of Diseases generally. Second Edition. Post 8vo, 10s.

The Prevention of Disease in Tropical and Sub-Tropical Campaigns. (Parkes Memorial Prize for 1886.) By Lieut.-Col. ANDREW DUNCAN, M.D., B.S. Lond., F.R.C.S., Indian Medical Service. 8vo, 12s. 6d.

A Commentary on the Diseases of India. By NORMAN CHEVERS, C.I.E., M.D., F.R.C.S., Deputy Surgeon-General H.M. Indian Army. 8vo, 24s.

Hooper's Physicians' Vade-Mecum. A Manual of the Principles and Practice of Physic. Tenth Edition. By W. A. GUY, F.R.C.P., F.R.S., and J. HARLEY, M.D., F.R.C.P. With 118 Engravings. Fcap. 8vo, 12s. 6d.

LONDON: 7, GREAT MARLBOROUGH STREET.

J. & A. CHURCHILL'S RECENT WORKS.

The Principles and Practice of Medicine. (Text-book.) By the late C. HILTON FAGGE, M.D., and P. H. PYE-SMITH, M.D., F.R.S., F.R.C.P., Physician to, and Lecturer on Medicine in, Guy's Hospital. Third Edition. 2 vols. 8vo, cloth, 40s.; Half Leather, 46s.

Manual of the Practice of Medicine. By FREDERICK TAYLOR, M.D., F.R.C.P., Physician to, and Lecturer on Medicine at, Guy's Hospital. Fifth Edition. 8vo, with Engravings, 16s.

The Practice of Medicine. By M. CHARTERIS, M.D., Professor of Therapeutics and Materia Medica in the University of Glasgow. Eighth Edition. Edited by F J. CHARTERIS, M.B., Ch.B. Crown 8vo, with Engravings on Copper and Wood, 10s.

A Dictionary of Practical Medicine. By various writers. Edited by JAS. KINGSTON FOWLER, M.A., M.D., F.R.C.P., Physician to Middlesex Hospital and the Hospital for Consumption. 8vo, cloth, 21s.; half calf, 25s.

A Text-Book of Bacteriology, for Medical Students and Practitioners. By G. M. STERNBERG, M.D., Surgeon-General, U.S. Army. With 9 Plates and 200 Figures in the Text. 8vo, 24s.

How to Examine the Chest: A Practical Guide for the use of Students. By SAMUEL WEST, M.D., F.R.C.P., Assistant Physician to St. Bartholomew's Hospital. Second Edition. With Engravings. Fcap. 8vo, 5s.

An Atlas of the Pathological Anatomy of the Lungs. By the late WILSON FOX, M.D., F.R.S., F.R.C.P., Physician to H.M. the Queen. With 45 Plates (mostly Coloured) and Engravings. 4to, half-bound in Calf, 70s.

By the same Author.

A Treatise on Diseases of the Lungs and Pleura. Edited by SIDNEY COUPLAND, M.D., F.R.C.P., Physician to Middlesex Hospital. Roy. 8vo, with Engravings; also Portrait and Memoir of the Author, 36s.

The Student's Guide to Diseases of the Chest. By VINCENT D. HARRIS, M.D. Lond., F.R.C.P., Physician to the City of London Hospital for Diseases of the Chest, Victoria Park. Fcap. 8vo, with 55 Illustrations (some Coloured), 7s. 6d.

Uric Acid as a Factor in the Causation of Disease. By ALEXANDER HAIG, M.D., F.R.C.P., Physician to the Metropolitan Hospital and the Royal Hospital for Children and Women. Fourth Edition. With 65 Illustrations, 8vo, 12s. 6d.

Text-Book of Medical Treatment (Diseases and Symptoms). By NESTOR I. C. TIRARD, M.D., F.R.C.P., Professor of the Principles and Practice of Medicine, King's College, London. 8vo, 15s.

Medical Diagnosis (Student's Guide Series). By SAMUEL FENWICK, M.D., F.R.C.P., and W. SOLTAU FENWICK, M.D., B.S. Eighth Edition. Crown 8vo, with 135 Engravings, 9s.

By the same Authors.

Outlines of Medical Treatment. Fourth Edition. Crown 8vo, with 35 Engravings, 10s.

Also.

Ulcer of the Stomach and Duodenum With 55 Illustrations. Royal 8vo, 10s. 6d.

Also, by Dr. SAMUEL FENWICK.

Clinical Lectures on Some Obscure Diseases of the Abdomen. Delivered at the London Hospital. 8vo, with Engravings, 7s. 6d.

Also.

The Saliva as a Test for Functional Diseases of the Liver. Crown 8vo, 2s.

The Liver. By LIONEL S. BEALE, M.B., F.R.S., Consulting Physician to King's College Hospital. With 24 Plates (85 Figures). 8vo, 5s.

By the same Author.

On Slight Ailments: And on Treating Disease. Fourth Edition. 8vo, 5s.

Myxœdema and the Thyroid Gland. By JOHN D. GIMLETTE, M.R.C.S., L.R.C.P. Crown 8vo, 5s.

The Blood: How to Examine and Diagnose its Diseases. By ALFRED C. COLES, M.D., B.Sc. With 6 Coloured Plates. 8vo, 10s. 6d.

The Physiology of the Carbohydrates; their Application as Food and Relation to Diabetes. By F. W. PAVY, M.D., LL.D., F.R.S., F.R.C.P., Consulting Physician to Guy's Hospital. Royal 8vo, with Plates and Engravings, 10s. 6d.

Medical Lectures and Essays. By Sir G. JOHNSON, M.D., F.R.C.P., F.R.S. 8vo, with 46 Engravings, 25s.

By the same Author.

An Essay on Asphyxia (Apnœa). 8vo, 3s.

Bronchial Asthma: Its Pathology and Treatment. By J. B. BERKART, M.D., late Physician to the City of London Hospital for Diseases of the Chest. Second Edition, with 7 Plates (35 Figures). 8vo, 10s. 6d.

LONDON: 7, GREAT MARLBOROUGH STREET.

Treatment of Some of the Forms of Valvular Disease of the Heart. By A. E. SANSOM, M.D., F.R.C.P., Physician to the London Hospital. Second Edition. Fcap. 8vo, with 26 Engravings, 4s. 6d.

The Schott Methods of the Treatment of Chronic Diseases of the Heart, with an account of the Nauheim Baths and of the Therapeutic Exercises. By W. BEZLY THORNE, M.D., M.R.C.P. Third Edition. 8vo, with Illustrations, 6s.

Guy's Hospital Reports. By the Medical and Surgical Staff. Vol. XXXVIII. Third Series. 8vo, 10s. 6d.

St. Thomas's Hospital Reports. By the Medical and Surgical Staff. Vol. XXVI. New Series. 8vo, 8s. 6d.

Westminster Hospital Reports. By the Medical and Surgical Staff. Vol. XI. 8vo, 8s.

Medical Ophthalmoscopy: A Manual and Atlas. By SIR WILLIAM R. GOWERS, M.D., F.R.C.P., F.R.S. Third Edition. Edited with the assistance of MARCUS GUNN, M.B., F.R.C.S., Surgeon to the Royal London Ophthalmic Hospital. With Coloured Plates and Woodcuts. 8vo, 16s.

By the same Author.

A Manual of Diseases of the Nervous System. Roy. 8vo.
Vol. I. Nerves and Spinal Cord. Third Edition by the Author and JAMES TAYLOR, M.D., F.R.C.P. With 192 Engravings, 15s.
Vol. II. Brain and Cranial Nerves: General and Functional Diseases. Second Edition. With 182 Engravings, 20s.

Also.

Clinical Lectures on Diseases of the Nervous System. 8vo, 7s. 6d.

Also.

Diagnosis of Diseases of the Brain. Second Edition. 8vo, with Engravings, 7s. 6d.

Also.

Syphilis and the Nervous System. Lettsomian Lectures for 1890. Delivered before the Medical Society of London. 8vo, 4s.

The Nervous System, Diseases of. By J. A. ORMEROD, M.D., F.R.C.P., Physician to the National Hospital for the Paralysed and Epileptic. With 66 Illustrations. Fcap. 8vo, 8s. 6d.

Text-Book of Nervous Diseases, for Students and Practitioners of Medicine. By CHARLES L. DANA, M.D., Professor of Nervous and Mental Diseases in Bellevue Hospital Medical College, New York. Fourth Edition, with 246 Illustrations. 8vo, 20s.

Handbook of the Diseases of the Nervous System. By JAMES ROSS, M.D., F.R.C.P., Professor of Medicine in the Victoria University, and Physician to the Royal Infirmary, Manchester. Roy. 8vo, with 184 Engravings, 18s.

Also.

Aphasia: Being a Contribution to the Subject of the Dissolution of Speech from Cerebral Disease. 8vo, with Engravings. 4s. 6d.

Diseases of the Nervous System. Lectures delivered at Guy's Hospital. By SIR SAMUEL WILKS, BART., M.D., F.R.S. Second Edition. 8vo, 18s.

Stammering: Its Causes, Treatment, and Cure. By A. G. BERNARD, M.R.C.S., L.R.C.P. Crown 8vo, 2s.

Secondary Degenerations of the Spinal Cord (Gulstonian Lectures, 1889). By HOWARD H. TOOTH, M.D., F.R.C.P., Assistant Physician to the National Hospital for the Paralysed and Epileptic. With Plates and Engravings. 8vo, 3s. 6d.

Diseases of the Nervous System. Clinical Lectures. By THOMAS BUZZARD, M.D., F.R.C.P., Physician to the National Hospital for the Paralysed and Epileptic. With Engravings, 8vo. 15s.

By the same Author.

Some Forms of Paralysis from Peripheral Neuritis: of Gouty, Alcoholic, Diphtheritic, and other origin. Crown 8vo, 5s.

Also.

On the Simulation of Hysteria by Organic Disease of the Nervous System. Crown 8vo, 4s. 6d.

Gout in its Clinical Aspects. By J. MORTIMER GRANVILLE, M.D. Crown 8vo, 6s.

Diseases of the Liver: With and without Jaundice By GEORGE HARLEY, M.D., F.R.C.P., F.R.S. 8vo, with 2 Plates and 36 Engravings, 21s.

Rheumatic Diseases, (Differentiation in). By HUGH LANE, Surgeon to the Royal Mineral Water Hospital, Bath, and Hon. Medical Officer to the Royal United Hospital, Bath. Second Edition, much Enlarged, with 8 Plates. Crown 8vo, 3s. 6d.

Diseases of the Abdomen, Comprising those of the Stomach and other parts of the Alimentary Canal, Œsophagus, Cæcum, Intestines, and Peritoneum. By S. O. HABERSHON, M.D., F.R.C.P. Fourth Edition. 8vo, with 5 Plates, 21s.

On Gallstones, or Cholelithiasis. By E. M. BROCKBANK, M.D. Vict., M.R.C P. Lond., Honorary Physician to the Ancoats Hospital, Manchester. Crown 8vo, 7s.

On the Relief of Excessive and Dangerous Tympanites by Puncturing the Abdomen. By JOHN W. OGLE, M.D., Consulting Physician to St. George's Hospital. 8vo, 5s. 6d.

Headaches : Their Nature, Causes, and Treatment. By W. H. DAY, M.D., Physician to the Samaritan Hospital. Fourth Edition. Crown 8vo, with Engravings, 7s. 6d.

A Handbook of Medical Climatology; embodying its Principles and Therapeutic Application, with Scientific Data of the Chief Health Resorts of the World. By S. EDWIN SOLLY, M.D., M.R.C.S., late President of the American Climatological Association. With Engravings and Coloured Plates. 8vo., 16s.

The Mineral Waters of France And its Wintering Stations (Medical Guide to). With a Special Map. By A. VINTRAS, M.D., Physician to the French Embassy, and to the French Hospital, London. Second Edition. Crown 8vo, 8s.

Illustrated Ambulance Lectures: To which is added a NURSING LECTURE. By JOHN M. H. MARTIN, M.D., F.R.C.S., Honorary Surgeon to the Blackburn Infirmary. Fourth Edition. Crown 8vo, with 60 Engravings, 2s.

Surgery: its Theory and Practice. By WILLIAM J. WALSHAM, F.R.C.S., Senior Assistant Surgeon to, and Lecturer on Anatomy at, St. Bartholomew's Hospital. Sixth Edition. Crown 8vo, with 410 Engravings, 12s. 6d.

A Synopsis of Surgery. By R. F. TOBIN, Surgeon to St. Vincent's Hospital, Dublin. Crown 8vo, interleaved, leather binding, 6s. 6d.

The Surgeon's Vade-Mecum : A Manual of Modern Surgery. By R. DRUITT, F.R.C.S. Twelfth Edition. By STANLEY BOYD, M.B., F.R.C.S. Crown 8vo, with 373 Engravings, 16s.

Operations on the Brain (A Guide to). By ALEC FRASER, Professor of Anatomy, Royal College of Surgeons in Ireland. Illustrated by 42 life-size Plates in Autotype, and 2 Woodcuts in the text. Folio, 63s.

The Operations of Surgery : Intended for Use on the Dead and Living Subject alike. By W. H. A. JACOBSON, M.A., M.B., M.Ch. Oxon., F.R.C.S., Assistant Surgeon to, and Lecturer on Anatomy at, Guy's Hospital. Third Edition. 8vo, with 401 Illustrations, 34s.

A Course of Operative Surgery. By CHRISTOPHER HEATH, Surgeon to University College Hospital. Second Edition. With 20 coloured Plates (180 figures) from Nature, by M. LÉVEILLÉ, and several Woodcuts. Large 8vo, 30s.

By the same Author.

The Student's Guide to Surgical Diagnosis. Second Edition. Fcap. 8vo, 6s. 6d.

Also.

Manual of Minor Surgery and Bandaging. For the use of House Surgeons, Dressers, and Junior Practitioners. Eleventh Edition. Fcap. 8vo, with 176 Engravings, 6s.

Also.

Injuries and Diseases of the Jaws. Fourth Edition. By HENRY PERCY DEAN, M.S., F.R.C.S., Assistant Surgeon to the London Hospital. 8vo, with 187 Wood Engravings, 14s.

Also.

Lectures on Certain Diseases of the Jaws. Delivered at the R.C.S., Eng., 1887. 8vo, with 64 Engravings, 2s. 6d.

Also.

Clinical Lectures on Surgical Subjects. Delivered in University College Hospital. Second Edition, Enlarged. Fcap. 8vo, with 27 Engravings, 6s.

Surgery. By C. W. MANSELL MOULLIN, M.A., M.D., Oxon., F.R.C.S., Surgeon and Lecturer on Physiology to the London Hospital. Large 8vo, with 497 Engravings, 34s.

The Practice of Surgery : A Manual. By THOMAS BRYANT, Consulting Surgeon to Guy's Hospital. Fourth Edition. 2 vols. crown 8vo, with 750 Engravings (many being coloured), and including 6 chromo plates, 32s.

Surgical Emergencies : Together with the Emergencies attendant on Parturition and the Treatment of Poisoning. By PAUL SWAIN, F.R.C.S., Surgeon to the South Devon and East Cornwall Hospital. Fifth Edition. Crown 8vo, with 149 Engravings, 6s.

Diseases of Bones and Joints. By CHARLES MACNAMARA, F.R.C.S., Surgeon to, and Lecturer on Surgery at, the Westminster Hospital. 8vo, with Plates and Engravings, 12s.

Abdominal Surgery. By J. GREIG SMITH, M.A., F.R.S.E. Sixth Edition. Edited by JAMES SWAIN, M.S., M.D. Lond., F.R.C.S. Eng., Assistant Surgeon to the Bristol Royal Infirmary, Professor of Surgery, University College, Bristol. Two Vols., 8vo, with 224 Engravings, 36s.

J. & A. CHURCHILL'S RECENT WORKS.   9

The Surgery of the Alimentary Canal. By ALFRED ERNEST MAYLARD, M.B. Lond. and B.S., Surgeon to the Victoria Infirmary, Glasgow. With 27 Swantype Plates and 89 Figures in the text, 8vo, 25s.

By the same Author.

A Student's Handbook of the Surgery of the Alimentary Canal. With 97 Illustrations. Crown 8vo, 8s. 6d.

Ovariotomy and Abdominal Surgery. By HARRISON CRIPPS, F.R.C.S., Surgical Staff, St. Bartholomew's Hospital. With numerous Plates, 8vo, 25s.

The Physiology of Death from Traumatic Fever: A Study in Abdominal Surgery. By JOHN D. MALCOLM, M.B., C.M., F.R.C.S.E., Surgeon to the Samaritan Free Hospital. 8vo, 3s. 6d.

Surgical Pathology and Morbid Anatomy. By ANTHONY A. BOWLBY, F.R.C.S., Assistant Surgeon to St. Bartholomew's Hospital. Third Edition. Crown 8vo, with 183 Engravings, 10s. 6d.

By the same Author.

Injuries and Diseases of Nerves and their Surgical Treatment. 8vo, with 20 Plates, 14s.

The Deformities of the Fingers and Toes. By WILLIAM ANDERSON, F.R.C.S., Surgeon to St. Thomas's Hospital. 8vo, with 18 Engravings, 6s.

Short Manual of Orthopædy. By HEATHER BIGG, F.R.C.S. Ed. Part I. Deformities and Deficiencies of the Head and Neck. 8vo, 2s. 6d.

The Human Foot: Its Form and Structure, Functions and Clothing. By THOMAS S. ELLIS, Consulting Surgeon to the Gloucester Infirmary. With 7 Plates and Engravings (50 Figures). 8vo, 7s. 6d.

Face and Foot Deformities. By FREDERICK CHURCHILL, C.M. 8vo, with Plates and Illustrations, 10s. 6d.

Royal London Ophthalmic Hospital Reports. By the Medical and Surgical Staff. Vol. XIV., Part 2. 8vo, 5s.

Ophthalmological Society of the United Kingdom. Transactions, Vol. XIX. 8vo, 12s. 6d.

Manual of Ophthalmic Surgery and Medicine. By W. H. H. JESSOP, M.A., F.R.C.S., Ophthalmic Surgeon to St. Bartholomew's Hospital. With 5 Coloured Plates and 110 Woodcuts, crown 8vo, 9s. 6d.

Nettleship's Diseases of the Eye: A Manual for Students. Sixth Edition, revised and edited by W. T. HOLMES SPICER, M.B., F.R.C.S., Ophthalmic Surgeon to the Metropolitan Hospital and the Victoria Hospital for Children. With 161 Engravings and a Coloured Plate illustrating Colour-Blindness, crown 8vo, 8s. 6d.

Diseases and Refraction of the Eye. By N. C. MACNAMARA, F.R.C.S., Surgeon to Westminster Hospital, and GUSTAVUS HARTRIDGE, F.R.C.S., Surgeon to the Royal Westminster Ophthalmic Hospital. Fifth Edition. Crown 8vo, with Plate, 156 Engravings, also Test-types, 10s. 6d.

Diseases of the Eye: a Practical Handbook for General Practitioners and Students. By CECIL EDWARD SHAW, M.D., M.Ch., Ophthalmic Surgeon to the Ulster Hospital for Children and Women, Belfast. With a Test-Card for Colour - Blindness. Crown 8vo, 3s. 6d.

On Diseases and Injuries of the Eye: A Course of Systematic and Clinical Lectures to Students and Medical Practitioners. By J. R. WOLFE, M.D., F.R.C.S.E., Lecturer on Ophthalmic Medicine and Surgery in Anderson's College, Glasgow. With 10 Coloured Plates and 157 Wood Engravings. 8vo, £1 1s.

Normal and Pathological Histology of the Human Eye and Eyelids. By C. FRED. POLLOCK, M.D., F.R.C.S. and F.R.S.E., Surgeon for Diseases of the Eye to Anderson's College Dispensary, Glasgow. Crown 8vo, with 100 Plates (230 drawings), 15s.

Convergent Strabismus, and its Treatment; an Essay. By EDWIN HOLTHOUSE, M.A., F.R.C.S., Surgeon to the Western Ophthalmic Hospital. 8vo, 6s.

Refraction of the Eye: A Manual for Students. By GUSTAVUS HARTRIDGE, F.R.C.S., Surgeon to the Royal Westminster Ophthalmic Hospital. Tenth Edition. Crown 8vo, with 104 Illustrations, also Test-types, &c., 6s.

By the same Author.

The Ophthalmoscope: A Manual for Students. Third Edition. Crown 8vo, with 68 Illustrations and 4 Plates. 4s. 6d.

Methods of Operating for Cataract and Secondary Impairments of Vision, with the results of 500 cases. By Major G. H. FINK, H.M. Indian Medical Service. Crown 8vo, with 15 Engravings, 5s.

LONDON: 7, GREAT MARLBOROUGH STREET.

Atlas of Ophthalmoscopy.
Composed of 12 Chromo-lithographic Plates (59 Figures drawn from nature) and Explanatory Text. By RICHARD LIEBREICH, M.R.C.S. Translated by H. ROSBOROUGH SWANZY, M.B. Third Edition, 4to, 40s.

Glaucoma:
Its Pathology and Treatment. By PRIESTLEY SMITH, Ophthalmic Surgeon to, and Clinical Lecturer on Ophthalmology at, the Queen's Hospital, Birmingham. 8vo, with 64 Engravings and 12 Zinco-photographs, 7s. 6d.

Eyestrain
(commonly called Asthenopia). By ERNEST CLARKE, M.D., B.S Lond., Surgeon to the Central London Ophthalmic Hospital, Surgeon and Ophthalmic Surgeon to the Miller Hospital. Second Edition. 8vo, with 22 Illustrations, 5s.

Diseases and Injuries of the Ear.
By Sir WILLIAM B. DALBY, F.R.C.S., M.B., Consulting Aural Surgeon to St. George's Hospital. Fourth Edition. Crown 8vo, with 8 Coloured Plates and 38 Wood Engravings. 10s. 6d.

By the same Author.

Short Contributions to Aural Surgery, between 1875 and 1896.
Third Edition. 8vo, with Engravings, 5s.

Diseases of the Ear,
Including the Anatomy and Physiology of the Organ, together with the Treatment of the Affections of the Nose and Pharynx, which conduce to Aural Disease. By T. MARK HOVELL, F.R.C.S.E., Aural Surgeon to the London Hospital, and Lecturer on Diseases of the Throat in the College, &c. 8vo, with 122 Engravings, 18s.

A System of Dental Surgery.
By Sir JOHN TOMES, F.R S., and C. S. TOMES, M.A., F.R.S. Fourth Edition. Post 8vo, with 289 Engravings, 16s.

Dental Anatomy, Human and Comparative:
A Manual. By CHARLES S. TOMES, M.A., F.R.S. Fifth Edition. Post 8vo, with 263 Engravings, 14s.

Dental Caries: an Investigation into its Cause and Prevention.
By J. SIM WALLACE, M.D., B.Sc., L.D.S.R.C.S. 8vo, 5s.

Dental Materia Medica, Pharmacology and Therapeutics.
By CHARLES W. GLASSINGTON, M.R.C.S., L.D.S. Edin.; Senior Dental Surgeon, Westminster Hospital; Dental Surgeon, National Dental Hospital; and Lecturer on Dental Materia Medica and Therapeutics to the College. Crown 8vo, 6s.

Dental Medicine:
A Manual of Dental Materia Medica and Therapeutics. By FERDINAND J. S. GORGAS, M.D., D.D.S., Professor of the Principles of Dental Science in the University of Maryland. Sixth Edition. 8vo, 18s.

A Manual of Dental Metallurgy.
By ERNEST A. SMITH, F.I.C., Assistant Instructor in Metallurgy, Royal College of Science, London. With 37 Illustrations. Crown 8vo, 6s. 6d.

A Practical Treatise on Mechanical Dentistry.
By JOSEPH RICHARDSON, M.D., D.D.S. Seventh Edition revised and Edited by GEORGE W. WARREN, D.D.S. Roy. 8vo, with 690 Engravings, 22s.

A Manual of Nitrous Oxide Anæsthesia,
for the use of Students and General Practitioners. By J. FREDERICK W. SILK, M.D. Lond., M.R.C.S., Anæsthetist to the Royal Free Hospital, Dental School of Guy's Hospital, and National Epileptic Hospital. 8vo, with 26 Engravings, 5s.

Skin Diseases of Children.
By GEO. H. FOX, M.D., Clinical Professor of Diseases of the Skin, College of Physicians and Surgeons, New York. With 12 Photogravure and Chromographic Plates, and 60 Illustrations in the Text, Roy. 8vo, 12s. 6d.

A Handbook on Leprosy.
By S. P. IMPEY, M.D., M.C., late Chief and Medical Superintendent, Robben Island Leper and Lunatic Asylums, Cape Colony. With 38 Plates and Map, 8vo, 12s.

Diseases of the Skin
(Introduction to the Study of). By P. H. PYE-SMITH, M.D., F.R.S., F.R.C.P., Physician to, and Lecturer on Medicine in, Guy's Hospital. Crown 8vo, with 26 Engravings. 7s. 6d.

A Manual of Diseases of the Skin,
with an Analysis of 20,000 consecutive Cases and a Formulary. By DUNCAN L. BULKLEY, M.D., New York. Fourth Edition, Roy. 16mo, 6s. 6d.

Cancerous Affections of the Skin.
(Epithelioma and Rodent Ulcer.) By GEORGE THIN, M.D. Post 8vo, with 8 Engravings, 5s.

By the same Author.

Pathology and Treatment of Ringworm.
8vo, with 21 Engravings, 5s.

LONDON: 7, GREAT MARLBOROUGH STREET.

The Operative Surgery of Malignant Disease. By HENRY T. BUTLIN, F.R.C.S., Surgeon to St. Bartholomew's Hospital. Second Edition, with 12 Engravings. 8vo, 14s.

By the same Author.
Malignant Disease (Sarcoma and Carcinoma) of the Larynx. 8vo, with 5 Engravings, 5s.

Also.
Sarcoma and Carcinoma: Their Pathology, Diagnosis, and Treatment. 8vo, with 4 Plates, 8s.

Cancers and the Cancer Process: a Treatise, Practical and Theoretic. By HERBERT L. SNOW, M.D., Surgeon to the Cancer Hospital, Brompton. 8vo, with 15 Lithographic Plates. 15s.

By the same Author.
The Re-appearance (Recurrence) of Cancer after apparent Extirpation. 8vo, 5s. 6d.

Also.
The Palliative Treatment of Incurable Cancer. Crown 8vo, 2s. 6d.

Ringworm and some other Scalp Affections: their Cause and Cure. By HAYDN BROWN, L.R.C.P.Ed. 8vo, 5s.

The Diagnosis and Treatment of Syphilis. By TOM ROBINSON, M.D., Physician to the Western Skin Hospital. Second Edition, Crown 8vo, 3s. 6d.

By the same Author.
The Diagnosis and Treatment of Eczema. Second Edition, Crown 8vo, 3s. 6d.

Also.
Illustrations of Diseases of the Skin and Syphilis, with Remarks. Fasc. I. with 3 Plates. Imp. 4to, 5s.

Selected Papers on Stone, Prostate, and other Urinary Disorders. By REGINALD HARRISON, F.R.C.S., Surgeon to St. Peter's Hospital. 8vo, with 15 Illustrations. 5s.

Chemistry of Urine; A Practical Guide to the Analytical Examination of Diabetic, Albuminous, and Gouty Urine. By ALFRED H. ALLEN, F.I.C., F.C.S. With Engravings, 8vo, 7s. 6d.

Clinical Chemistry of Urine (Outlines of the). By C. A. MAC MUNN, M.A., M.D. 8vo, with 64 Engravings and Plate of Spectra, 9s.

BY SIR HENRY THOMPSON, Bart., F.R.C.S.
Diseases of the Urinary Organs. Clinical Lectures. Eighth Edition. 8vo, with 121 Engravings, 10s. 6d.

Practical Lithotomy and Lithotrity; or, An Inquiry into the Best Modes of Removing Stone from the Bladder. Third Edition. 8vo, with 87 Engravings, 10s.

The Preventive Treatment of Calculous Disease, and the Use of Solvent Remedies. Third Edition. Crown 8vo, 2s. 6d.

Tumours of the Bladder: Their Nature, Symptoms, and Surgical Treatment. 8vo, with numerous Illustrations, 5s.

Stricture of the Urethra, and Urinary Fistulæ: their Pathology and Treatment. Fourth Edition. 8vo, with 74 Engravings, 6s.

The Suprapubic Operation of Opening the Bladder for the Stone and for Tumours. 8vo, with 14 Engravings, 3s. 6d.

The Clinical Examination of Urine, with an Atlas of Urinary Deposits. By LINDLEY SCOTT, M.A., M.D. With 41 Original Plates (mostly in Colours). Crown 4to, 15s.

Electric Illumination of the Bladder and Urethra, as a Means of Diagnosis of Obscure Vesico-Urethral Diseases. By E. HURRY FENWICK, F.R.C.S., Surgeon to London Hospital and St. Peter's Hospital for Stone. Second Edition. 8vo, with 54 Engravings, 6s. 6d.

By the same Author.
Tumours of the Urinary Bladder. The Jacksonian Prize Essay of 1887, rewritten with 200 additional cases, in four Fasciculi. Fas. I. Royal 8vo, 5s.

Also.
The Cardinal Symptoms of Urinary Diseases: their Diagnostic Significance and Treatment. 8vo, with 36 Illustrations. 8s. 6d.

Atlas of Electric Cystoscopy. By Dr. EMIL BURCKHARDT, late of the Surgical Clinique of the University of Bâle, and E. HURRY FENWICK, F.R.C.S., Surgeon to the London Hospital and St. Peter's Hospital. Royal 8vo, with 34 Coloured Plates, embracing 83 Figures. 21s.

LONDON: 7, GREAT MARLBOROUGH STREET.

Urinary and Renal Derangements and Calculous Disorders. By LIONEL S. BEALE, F.R.C.P., F.R.S., Consulting Physician to King's College Hospital. 8vo, 5s.

Male Organs of Generation (Diseases of). By W. H. A. JACOBSON, M.Ch. Oxon., F.R.C.S., Assistant Surgeon to Guy's Hospital. 8vo, with 88 Engravings. 22s.

The Surgical Diseases of the Genito - Urinary Organs, including Syphilis. By E. L. KEYES, M.D., Professor in Bellevue Hospital Medical College, New York (a revision of VAN BUREN and KEYES' Text-book). Roy. 8vo, with 114 Engravings, 21s.

Diseases of the Rectum and Anus. By ALFRED COOPER, F.R.C.S., Senior Surgeon to the St. Mark's Hospital for Fistula ; and F. SWINFORD EDWARDS, F.R.C.S., Senior Assistant Surgeon to St. Mark's Hospital. Second Edition, with Illustrations. 8vo, 12s.

Diseases of the Rectum and Anus. By HARRISON CRIPPS, F.R.C.S., Assistant Surgeon to St. Bartholomew's Hospital, &c. Second Edition. 8vo, with 13 Lithographic Plates and numerous Wood Engravings, 12s. 6d.

By the same Author.

Cancer of the Rectum. Especially considered with regard to its Surgical Treatment. Jacksonian Prize Essay. 8vo, with 13 Plates and several Wood Engravings, 6s.

Also.

The Passage of Air and Fæces from the Urethra. 8vo, 3s. 6d.

Syphilis. By ALFRED COOPER, F.R.C.S., Consulting Surgeon to the West London and the Lock Hospitals. Second Edition. Edited by EDWARD COTTERELL, F.R.C.S., Surgeon (out-patients) to the London Lock Hospital. 8vo, with 24 Full-page Plates (12 coloured), 18s.

On Maternal Syphilis, including the presence and recognition of Syphilitic Pelvic Disease in Women. By JOHN A. SHAW-MACKENZIE, M.D. 8vo, with Coloured Plates, 10s. 6d.

A Medical Vocabulary : An Explanation of all Terms and Phrases used in the various Departments of Medical Science and Practice, their Derivation, Meaning, Application, and Pronunciation. By R. G. MAYNE, M.D., LL.D. Sixth Edition by W. W. WAGSTAFFE, B.A., F.R.C.S. Crown 8vo, 10s. 6d.

A Short Dictionary of Medical Terms. Being an Abridgment of Mayne's Vocabulary. 64mo, 2s. 6d.

Dunglison's Dictionary of Medical Science : Containing a full Explanation of its various Subjects and Terms, with their Pronunciation, Accentuation, and Derivation. Twenty-first Edition. By RICHARD J. DUNGLISON, A.M., M.D. Royal 8vo, 30s.

Terminologia Medica Polyglotta : a Concise International Dictionary of Medical Terms (French, Latin, English, German, Italian, Spanish, and Russian). By THEODORE MAXWELL, M.D., B.Sc., F.R.C.S. Edin. Royal 8vo, 16s.

A German-English Dictionary of Medical Terms. By FREDERICK TREVES, F.R.C.S., Surgeon to the London Hospital ; and HUGO LANG, B.A. Crown 8vo, half-Persian calf, 12s.

A Handbook of Physics and Chemistry, adapted to the requirements of the first Examination of the Conjoint Board and for general use. By HERBERT E. CORBIN, B.Sc. Lond., and ARCHIBALD M. STEWART, B.Sc. Lond. With 120 Illustrations. Crown 8vo, 6s. 6d.

A Manual of Chemistry, Theoretical and Practical. By WILLIAM A. TILDEN, D.Sc., F.R.S., Professor of Chemistry in the Royal College of Science, London ; Examiner in Chemistry to the Department of Science and Art. With 2 Plates and 143 Woodcuts, crown 8vo, 10s.

Chemistry, Inorganic and Organic. With Experiments. By CHARLES L. BLOXAM. Eighth Edition, by JOHN MILLAR THOMSON, F.R.S., Professor of Chemistry in King's College, London, and ARTHUR G. BLOXAM, Head of the Chemistry Department, The Goldsmiths' Institute, New Cross. 8vo, with nearly 300 Illustrations, 18s. 6d.

By the same Author.

Laboratory Teaching; Or, Progressive Exercises in Practical Chemistry. Sixth Edition. By ARTHUR G. BLOXAM. Crown 8vo, with 80 Engravings, 6s. 6d.

Watts' Organic Chemistry. Edited by WILLIAM A. TILDEN, D.Sc., F.R.S., Professor of Chemistry, Royal College of Science, London. Second Edition. Crown 8vo, with Engravings, 10s.

Practical Chemistry
And Qualitative Analysis. By FRANK CLOWES, D.Sc. Lond., Emeritus Professor of Chemistry in the University College, Nottingham. Seventh Edition. Post 8vo, with 101 Engravings and Frontispiece, 8s. 6d.

Quantitative Analysis.
By FRANK CLOWES, D.Sc. Lond., Emeritus Professor of Chemistry in the University College, Nottingham, and J. BERNARD COLEMAN, Assoc. R. C. Sci. Dublin ; Professor of Chemistry, South-West London Polytechnic. Fifth Edition. Post 8vo, with 117 Engravings, 10s.
By the same Authors.

Elementary Quantitative Analysis. Post 8vo, with 62 Engravings, 4s. 6d.
Also,

Elementary Practical Chemistry and Qualitative Analysis. Third Edition. Post 8vo, with 68 Engravings, 3s. 6d.

Qualitative Analysis.
By R. FRESENIUS. Translated by CHARLES E. GROVES, F.R.S. Tenth Edition. 8vo, with Coloured Plate of Spectra and 46 Engravings, 15s.
By the same Author.

Quantitative Analysis.
Seventh Edition.
Vol. I., Translated by A. VACHER. 8vo, with 106 Engravings, 15s.
Vol. II., Parts 1 to 5, Translated by C. E. GROVES, F.R.S. 8vo, with Engravings, 2s. 6d. each.

Inorganic Chemistry.
By SIR EDWARD FRANKLAND, K.C.B., D.C.L., LL.D., F.R.S., and FRANCIS R. JAPP, M.A., Ph.D., F.I.C., F.R.S., Professor of Chemistry in the University of Aberdeen. 8vo, with numerous Illustrations on Stone and Wood, 24s.

Inorganic Chemistry
(A System of). By WILLIAM RAMSAY, Ph.D., F.R.S., Professor of Chemistry in University College, London. 8vo, with Engravings, 15s.
By the same Author.

Elementary Systematic Chemistry for the Use of Schools and Colleges. With Engravings. Crown 8vo, 4s. 6d. ; Interleaved, 5s. 6d.

Valentin's Practical Chemistry and Qualitative and Quantitative Analysis. Edited by W. R. HODGKINSON, Ph.D., F.R.S.E., Professor of Chemistry and Physics in the Royal Military Academy, and Artillery College, Woolwich. Ninth Edition. 8vo, with Engravings and Map of Spectra, 9s. [The Tables separately, 2s. 6d.]

Practical Chemistry, Part I.
Qualitative Exercises and Analytical Tables for Students. By J. CAMPBELL BROWN, Professor of Chemistry in Victoria University and University College, Liverpool. Fourth Edition. 8vo, 2s. 6d.

The Analyst's Laboratory Companion : a Collection of Tables and Data for Chemists and Students. By ALFRED E. JOHNSON, A.R.C.S.I., F.I.C. Second Edition, enlarged, crown 8vo., cloth, 5s., leather, 6s. 6d.

Volumetric Analysis :
Or the Quantitative Estimation of Chemical Substances by Measure, applied to Liquids, Solids, and Gases. By FRANCIS SUTTON, F.C.S., F.I.C. Seventh Edition. 8vo, with 112 Engravings, 18s. 6d.

Commercial Organic Analysis :
a Treatise on the Properties, Modes of Assaying, Proximate Analytical Examination, &c., of the various Organic Chemicals and Products employed in the Arts, Manufactures, Medicine, &c. By ALFRED H. ALLEN, F.I.C., F.C.S. 8vo.
Vol. I.—Introduction, Alcohols, Neutral Alcoholic Derivatives, &c., Ethers, Vegetable Acids, Starch and its Isomers, Sugars, &c. Third Edition. 18s.

Vol. II.—Part I. Fixed Oils and Fats, Glycerin, Nitro - Glycerin, Dynamites and Smokeless Powders, Wool-Fats, Dégras, etc. Third Edition. 14s.

Vol. II.—Part II. Hydrocarbons, Petroleum and Coal-Tar Products, Asphalt, Phenols and Creosotes, &c. Third Edition. 14s.

Vol. II.—Part III. Terpenes, Essential Oils, Resins and Camphors, Acid Derivatives of Phenols, Aromatic Acids. [*In preparation.*

Vol. III.—Part I. Dyes and Colouring Matters. Third Edition.
[*Nearly ready.*

Vol. III.—Part II. The Anilines and Ammonium Bases, Hydrazines and Derivatives, Bases from Tar, The Antipyretics, &c., Vegetable Alkaloids, Tea, Coffee, Cocoa, Kola, Cocaine, Opium, &c. Second Edition. 18s.

Vol. III.—Part III. Vegetable Alkaloids, Non-Basic Vegetable Bitter Principles, Animal Bases, Animal Acids, Cyanogen and its Derivatives, &c. Second Edition. 16s.

Vol. IV.—The Proteïds and Albuminous Principles, Proteoïds or Albuminoïds. Second Edition. 18s.

LONDON: 7, GREAT MARLBOROUGH STREET.

J. & A. CHURCHILL'S RECENT WORKS.

Cooley's Cyclopædia
of Practical Receipts, and Collateral Information in the Arts, Manufactures, Professions, and Trades: Including Medicine, Pharmacy, Hygiene and Domestic Economy. Seventh Edition, by W. NORTH, M.A. Camb., F.C.S. 2 Vols., Roy. 8vo, with 371 Engravings, 42s.

Chemical Technology:
A Manual. By RUDOLF VON WAGNER. Translated and Edited by SIR WILLIAM CROOKES, F.R.S., from the Thirteenth Enlarged German Edition as remodelled by Dr. FERDINAND FISCHER. 8vo, with 596 Engravings, 32s.

Chemical Technology;
Or, Chemistry in its Applications to Arts and Manufactures. Edited by CHARLES E. GROVES, F.R.S., and WILLIAM THORP, B.Sc.
Vol. I.—FUEL AND ITS APPLICATIONS. By E. J. MILLS, D.Sc., F.R.S., and F. J. ROWAN, C.E. Royal 8vo, with 606 Engravings, 30s.
Vol. II.—LIGHTING BY CANDLES AND OIL. By W. Y. DENT, J. MCARTHUR, L. FIELD and F. A. FIELD, BOVERTON REDWOOD, and D. A. LOUIS. Royal 8vo, with 358 Engravings and Map, 20s.
Vol. III.—GAS AND ELECTRICITY. [*In the press.*]

Technological Handbooks.
EDITED BY JOHN GARDNER, F.I.C., F.C.S., and JAMES CAMERON, F.I.C.
BREWING, DISTILLING, AND WINE MANUFACTURE. Crown 8vo, with Engravings, 6s. 6d.
BLEACHING, DYEING, AND CALICO PRINTING. With Formulæ. Crown 8vo, with Engravings, 5s.
OILS, RESINS, AND VARNISHES. Crown 8vo, with Engravings. 7s. 6d.
SOAPS AND CANDLES. Crown 8vo, with 54 Engravings, 7s.

Chemistry an Exact Mechanical Philosophy. By FRED. G. EDWARDS. Illustrated. 8vo, 3s. 6d.

The Microscope and its Revelations. By the late WILLIAM B. CARPENTER, C.B., M.D., LL.D., F.R.S. Seventh Edition, by the Rev. W. H. DALLINGER, LL.D., F.R.S. With 21 Plates and 800 Wood Engravings. 8vo, 26s. Half Calf, 30s.

The Quarterly Journal of Microscopical Science. Edited by E. RAY LANKESTER, M.A., LL.D., F.R.S.; with the co-operation of ADAM SEDGWICK, M.A., F.R.S., W. F. R. WELDON, M.A., F.R.S., and SYDNEY J. HICKSON, M.A., F.R.S. Each Number, 10s.

Encyclopædia Medica, under the general editorship of CHALMERS WATSON, M.B., M.R.C.P.E. About 12 Volumes, 20s. each. Vols. I. to IV. now ready.

Methods and Formulæ
Used in the Preparation of Animal and Vegetable Tissues for Microscopical Examination, including the Staining of Bacteria. By PETER WYATT SQUIRE, F.L.S. Crown 8vo, 3s. 6d.

The Microtomist's Vade-Mecum:
A Handbook of the Methods of Microscopic Anatomy. By ARTHUR BOLLES LEE, Assistant in the Russian Laboratory of Zoology at Villefranche-sur-mer (Nice). Fourth Edition. 8vo, 15s.

Photo-Micrography
(Guide to the Science of). By EDWARD C. BOUSFIELD, L.R.C.P. Lond. 8vo, with 34 Engravings and Frontispiece, 6s.

An Introduction to Physical Measurements, with Appendices on Absolute Electrical Measurements, &c. By Dr. F. KOHLRAUSCH, Professor at the University of Strassburg. Third Edition, translated from the Seventh German Edition, by THOMAS HUTCHINSON WALLER, B.A., B.Sc., and HENRY RICHARDSON PROCTER, F.I.C., F.C.S. 8vo, with 91 Illustrations, 12s. 6d.

Tuson's Veterinary Pharmacopœia, including the Outlines of Materia Medica and Therapeutics. Fifth Edition. Edited by JAMES BAYNE, F.C.S., Professor of Chemistry and Toxicology in the Royal Veterinary College. Crown 8vo, 7s. 6d.

The Veterinarian's Pocket Remembrancer: being Concise Directions for the Treatment of Urgent or Rare Cases, embracing Semeiology, Diagnosis, Prognosis, Surgery, Therapeutics, Toxicology, Detection of Poisons by their Appropriate Tests, Hygiene, &c. By GEORGE ARMATAGE, M.R.C.V.S. Second Edition. Post 8vo, 3s.

Chauveau's Comparative Anatomy of the Domesticated Animals, revised and enlarged, with the Co-operation of S. ARLOING, Director of the Lyons Veterinary School. Edited by GEO. FLEMING, C.B., LL.D., F.R.C.V.S., late Principal Veterinary Surgeon of the British Army. Second English Edition. 8vo, with 585 Engravings, 31s. 6d.

Human Nature, its Principles and the Principles of Physiognomy. By PHYSICIST. Part I., Imp. 16mo, 2s. Part II. (completing the work), 2s. 6d.

LONDON: 7, GREAT MARLBOROUGH STREET.

INDEX TO J. & A. CHURCHILL'S LIST.

Allen's Chemistry of Urine, 11
—— Commercial Organic Analysis, 13
Anderson's Deformities of Fingers and Toes, 9
Armatage's Veterinary Pocket Remembrancer, 14
Barnes (R.) on Obstetric Operations, 3
—— on Diseases of Women, 3
Beale (L. S.) on Liver, 6
—— Slight Ailments, 6
—— Urinary and Renal Derangements, 12
Beale (P. T. B.) on Elementary Biology, 2
Beasley's Book of Prescriptions, 5
—— Druggists' General Receipt Book, 5
—— Pharmaceutical Formulary, 5
Bell on Sterility, 4
Bellamy's Surgical Anatomy, 1
Bentley and Trimen's Medicinal Plants, 5
Bentley's Systematic Botany, 5
Berkart's Bronchial Asthma, 6
Bernard on Stammering, 7
Bigg's Short Manual of Orthopædy, 9
Birch's Practical Physiology, 4
Bloxam's Chemistry, 12
—— Laboratory Teaching, 12
Bousfield's Photo-Micrography, 14
Bowlby's Injuries and Diseases of Nerves, 9
—— Surgical Pathology and Morbid Anatomy, 9
Brockbank on Gallstones, 8
Brown's (Haydn) Midwifery, 3
—— —— Ringworm, 11
—— (Campbell) Practical Chemistry, 13
Bryant's Practice of Surgery, 8
Bulkley on Skin Diseases, 10
Burckhardt's (E.) and Fenwick's (E. H.) Atlas of Cystoscopy, 11
Burdett's Hospitals and Asylums of the World, 2
Butler-Smythe's Ovariotomies, 4
Butlin's Operative Surgery of Malignant Disease, 11
—— Malignant Disease of the Larynx, 11
—— Sarcoma and Carcinoma, 11
Buzzard's Diseases of the Nervous System, 7
—— Peripheral Neuritis, 7
—— Simulation of Hysteria, 7
Cameron's Oils, Resins, and Varnishes, 14
—— Soaps and Candles, 14
Carpenter and Dallinger on the Microscope, 14
Cautley's Infant Feeding, 4
Charteris' Practice of Medicine, 6
Chauveau's Comparative Anatomy, 14
Chevers' Diseases of India, 5
Churchill's Face and Foot Deformities, 9
Clarke's Eyestrain, 10
Clouston's Lectures on Mental Diseases, 3
Clowes and Coleman's Quantitative Analysis, 13
—— —— Elmntry Practical Chemistry, 13
Clowes' Practical Chemistry, 13
Coles on Blood, 6
Cooley's Cyclopædia of Practical Receipts, 14
Cooper on Syphilis, 12
Cooper and Edwards' Diseases of the Rectum, 12
Corbin and Stewart's Physics and Chemistry, 12
Cripps' (H.) Ovariotomy and Abdominal Surgery, 9
—— Cancer of the Rectum, 12
—— Diseases of the Rectum and Anus, 12
—— Air and Fæces in Urethra, 12
Cripps' (R. A.) Galenic Pharmacy, 4
Cuff's Lectures to Nurses, 4
Cullingworth's Short Manual for Monthly Nurses, 4
Dalby's Diseases and Injuries of the Ear, 10
—— Short Contributions, 10
Dana on Nervous Diseases, 7
Day on Headaches, 8
Domville's Manual for Nurses, 4
Doran's Gynæcological Operations, 3
Druitt's Surgeon's Vade-Mecum, 8
Duncan (A.), on Prevention of Disease in Tropics, 5
Dunglison's Dictionary of Medical Science, 12
Edwards' Chemistry, 14
Ellis's (T. S.) Human Foot, 9
Encyclopædia Medica, 4
Fagge's Principles and Practice of Medicine, 6
Fayrer's Climate and Fevers of India, 5
—— Natural History, &c., of Cholera, 5
Fenwick (E. H.), Electric Illumination of Bladder, 11
—— Tumours of Urinary Bladder, 11
—— Symptoms of Urinary Diseases, 11
Fenwick's (S.) Medical Diagnosis, 6

Fenwick's (S.) Obscure Diseases of the Abdomen, 6
—— Outlines of Medical Treatment, 6
—— Ulcer of the Stomach and Duodenum, 6
—— The Saliva as a Test, 6
Fink's Operating for Cataract, 9
Fowler's Dictionary of Practical Medicine, 6
Fox (G. H.) on Skin Diseases of Children, 10
Fox (Wilson), Atlas of Pathological Anatomy of Lungs, 6
—— Treatise on Diseases of the Lungs 6
Frankland and Japp's Inorganic Chemistry, 13
Fraser's Operations on the Brain, 8
Fresenius' Qualitative Analysis, 13
—— Quantitative Analysis, 13
Galabin's Diseases of Women, 3
—— Manual of Midwifery, 3
Gardner's Bleaching, Dyeing, and Calico Printing, 14
—— Brewing, Distilling, and Wine Manuf., 14
Gimlette on Myxœdema, 6
Glassington's Dental Materia Medica, 10
Godlee's Atlas of Human Anatomy, 1
Goodhart's Diseases of Children, 4
Gorgas' Dental Medicine, 10
Gowers' Diagnosis of Diseases of the Brain, 7
—— Manual of Diseases of Nervous System, 7
—— Clinical Lectures, 7
—— Medical Ophthalmoscope, 7
—— Syphilis and the Nervous System, 7
Granville on Gout, 7
Green's Manual of Botany, 5
Greenish's Materia Medica, 4
Groves' and Thorp's Chemical Technology, 14
Guy's Hospital Reports, 7
Habershon's Diseases of the Abdomen, 7
Haig's Uric Acid, 6
—— Diet and Food, 2
Harley on Diseases of the Liver, 7
Harris's (V. D.) Diseases of Chest, 6
Harrison's Urinary Organs, 11
Hartridge's Refraction of the Eye, 9
—— Ophthalmoscope, 9
Hawthorne's Galenical Preparations of B.P., 5
Heath's Certain Diseases of the Jaws, 8
—— Clinical Lectures on Surgical Subjects, 8
—— Injuries and Diseases of the Jaws, 8
—— Minor Surgery and Bandaging, 8
—— Operative Surgery, 8
—— Practical Anatomy, 1
—— Surgical Diagnosis, 8
Hedley's Therapeutic Electricity, 5
Hellier's Notes on Gynæcological Nursing, 4
Hewlett's Bacteriology, 3
Hill on Cerebral Circulation, 7
Hirschfeld's Atlas of Central Nervous System, 2
Holden's Human Osteology, 1
—— Landmarks, 1
Holthouse on Strabismus, 9
Hooper's Physicians' Vade-Mecum, 5
Hovell's Diseases of the Ear, 10
Human Nature and Physiognomy, 14
Hyslop's Mental Physiology, 3
Impey on Leprosy, 11
Ireland on Mental Affections of Children, 3
Jacobson's Male Organs of Generation, 12
—— Operations of Surgery, 8
Jellett's Practice of Midwifery, 3
Jessop's Ophthalmic Surgery and Medicine, 9
Johnson's (Sir G.) Asphyxia, 6
—— Medical Lectures and Essays, 6
—— (A. E.) Analyst's Companion, 13
Journal of Mental Science
Kellogg on Mental Diseases, 3
Kelynack's Pathologist's Handbook, 1
Keyes' Genito-Urinary Organs and Syphilis, 12
Kohlrausch's Physical Measurements, 14
Lancereaux's Atlas of Pathological Anatomy, 2
Lane's Rheumatic Diseases, 7
Langdon-Down's Mental Affections of Childhood, 3
Lazarus-Barlow's General Pathology, 1
Lee's Microtomist's Vade Mecum, 14
Lescher's Recent Materia Medica, 4
Lewis (Bevan) on the Human Brain, 2
Liebreich (O.) on Borax and Boracic Acid, 2
Liebreich's (R). Atlas of Ophthalmoscopy, 10
Lucas's Practical Pharmacy, 4
MacMunn's Clinical Chemistry of Urine, 11
Macnamara's Diseases and Refraction of the Eye, 9
(Continued on the next page

LONDON: 7, GREAT MARLBOROUGH STREET.

INDEX TO J. & A. CHURCHILL'S LIST—*continued*.

Macnamara's Diseases of Bones and Joints, 8
McNeill's Epidemics and Isolation Hospitals, 2
Malcolm's Physiology of Death, 9
Marcet on Respiration, 1
Martin's Ambulance Lectures, 8
Maxwell's Terminologia Medica Polyglotta, 12
Maylard's Surgery of Alimentary Canal, 9
Mayne's Medical Vocabulary, 12
Microscopical Journal, 14
Mills and Rowan's Fuel and its Applications, 14
Moore's (N.) Pathological Anatomy of Diseases, 1
Moore's (Sir W. J.) Family Medicine for India, 5
——— Manual of the Diseases of India, 5
Morris's Human Anatomy, 1
——— Anatomy of Joints, 1
Moullin's (Mansell) Surgery, 8
Nettleship's Diseases of the Eye, 9
Notter and Firth's Hygiene, 2
Ogle on Tympanites, 8
Oliver's Abdominal Tumours, 3
——— Diseases of Women, 3
Ophthalmic (Royal London) Hospital Reports, 9
Ophthalmological Society's Transactions, 9
Ormerod's Diseases of the Nervous System, 7
Parkes' (E.A.) Practical Hygiene, 2
Parkes' (L.C.) Elements of Health, 2
Pavy's Carbohydrates, 6
Pereira's Selecta è Prescriptis, 5
Phillips' Materia Medica and Therapeutics, 4
Pitt-Lewis's Insane and the Law, 3
Pollock's Histology of the Eye and Eyelids, 9
Proctor's Practical Pharmacy, 4
Pye-Smith's Diseases of the Skin, 10
Ramsay's Elementary Systematic Chemistry, 13
——— Inorganic Chemistry, 13
Richardson's Mechanical Dentistry, 10
Richmond's Antiseptic Principles for Nurses, 4
Roberts' (D. Lloyd) Practice of Midwifery, 3
Robinson's (Tom) Eczema, 11
——— Illustrations of Skin Diseases, 11
——— Syphilis, 11
Ross's Aphasia, 7
——— Diseases of the Nervous System, 7
St. Thomas's Hospital Reports, 7
Sansom's Valvular Disease of the Heart, 7
Scott's Atlas of Urinary Deposits, 11
Shaw's Diseases of the Eye, 9
Shaw-Mackenzie on Maternal Syphilis, 12
Short Dictionary of Medical Terms, 12
Silk's Manual of Nitrous Oxide, 10
Smith's (Ernest A.) Dental Metallurgy, 10
Smith's (Eustace) Clinical Studies, 4
——— Disease in Children, 4
——— Wasting Diseases of Infants and Children, 4
Smith's (J. Greig) Abdominal Surgery, 8
Smith's (Priestley) Glaucoma, 10

Snow's Cancer and the Cancer Process, 11
——— Palliative Treatment of Cancer, 11
——— Reappearance of Cancer, 11
Solly's Medical Climatology, 8
Southall's Materia Medica, 5
Squire's (P.) Companion to the Pharmacopœia, 4
——— London Hospitals Pharmacopœias, 4
——— Methods and Formulæ, 14
Starling's Elements of Human Physiology, 2
Sternberg's Bacteriology, 5
Stevenson and Murphy's Hygiene, 2
Suttou's (J. B.), General Pathology, 1
Sutton's (F.) Volumetric Analysis, 13
Swain's Surgical Emergencies, 8
Swayne's Obstetric Aphorisms, 3
Taylor's (A. S.) Medical Jurisprudence, 2
Taylor's (F.) Practice of Medicine, 6
Thin's Cancerous Affections of the Skin, 10
——— Pathology and Treatment of Ringworm, 10
——— on Psilosis or "Sprue," 5
Thomas's Diseases of Women, 3
Thompson's (Sir H.) Calculous Disease, 11
——— Diseases of the Urinary Organs, 11
——— Lithotomy and Lithotrity, 11
——— Stricture of the Urethra, 11
——— Suprapubic Operation, 11
——— Tumours of the Bladder, 11
Thorne's Diseases of the Heart, 7
Thresh's Water Analysis, 2
Tilden's Manual of Chemistry, 12
Tirard's Medical Treatment, 6
Tobin's Surgery, 8
Tomes' (C. S.) Dental Anatomy, 10
Tomes' (J. and C. S.) Dental Surgery, 10
Tooth's Spinal Cord, 7
Treves and Lang's German-English Dictionary, 12
Tuke's Dictionary of Psychological Medicine, 3
Tuson's Veterinary Pharmacopœia, 14
Valentin and Hodgkinson's Practical Chemistry, 13
Vintras on the Mineral Waters, &c., of France, 8
Wagner's Chemical Technology, 14
Wallace on Dental Caries, 10
Walsham's Surgery: its Theory and Practice, 8
Waring's Indian Bazaar Medicines, 5
——— Practical Therapeutics, 5
Watts' Organic Chemistry, 12
West's (S.) How to Examine the Chest, 6
Westminster Hospital Reports, 7
White's (Hale) Materia Medica, Pharmacy, &c., 4
Wilks' Diseases of the Nervous System, 7
Wilson's (Sir E.) Anatomists' Vade-Mecum, 1
Wilson's (G.) Handbook of Hygiene, 2
Wolfe's Diseases and Injuries of the Eye, 9
Wynter and Wethered's Practical Pathology, 1
Year-Book of Pharmacy, 5
Yeo's (G. F.) Manual of Physiology, 2

N.B.—J. & A. Churchill's larger Catalogue of about 600 *works on Anatomy Physiology, Hygiene, Midwifery, Materia Medica, Medicine, Surgery, Chemistry, Botany, &c. &c., with a complete Index to their Subjects, for easy reference, will be forwarded post free on application.*

AMERICA.—*J. & A. Churchill being in constant communication with various publishing houses in America are able to conduct negotiations favourable to English Authors.*

LONDON: 7, GREAT MARLBOROUGH STREET.

www.ingramcontent.com/pod-product-compliance
Lightning Source LLC
Chambersburg PA
CBHW020255170426
43202CB00008B/379